Applied Mathematical Sciences | Volume 14

T. Yoshizawa

Stability Theory and the Existence of Periodic Solutions and Almost Periodic Solutions

Springer-Verlag New York · Heidelberg · Berlin
1975

T. Yoshizawa

Mathematical Institute

Tohoku University

Sendai, Japan

AMS Classifications: 34D20, 34C25

Library of Congress Cataloging in Publication Data

Yoshizawa, Taro, 1919-
  Stability theory and the existence of periodic
solutions and almost periodic solutions.

  (Applied mathematical sciences; v. 14)
  Bibliography: p.
  Includes index.
  1. Differential equations—Numerical solutions.
2. Stability. 3. Almost periodic functions.
I. Title. II. Series.
QA1.A647 vol. 14 [QA372] 510′.8s [515′.35] 74-28140

ISBN-13: 978-0-387-90112-1   e-ISBN-13: 978-1-4612-6376-0
DOI: 10.1007/978-1-4612-6376-0

PREFACE

Since there are several excellent books on stability theory, the author selected some recent topics in stability theory which are related to existence theorems for periodic solutions and for almost periodic solutions. The author hopes that these notes will also serve as an introduction to stability theory. These notes contain stability theory by Liapunov's second method and somewhat extended discussion of stability properties in almost periodic systems, and the existence of a periodic solution in a periodic system is discussed in connection with the boundedness of solutions, and the existence of an almost periodic solution in an almost periodic system is considered in connection with some stability property of a bounded solution. In the theory of almost periodic systems, one has to consider almost periodic functions depending on parameters, but most of text books on almost periodic functions do not contain this case. Therefore, as mathematical preliminaries, the first chapter is intended to provide a guide for some properties of almost periodic functions with parameters as well as for properties of asymptotically almost periodic functions.

These notes originate from a seminar on stability theory given by the author at the Mathematics Department of Michigan State University during the academic year 1972-1973. The author is very grateful to Professor Pui-Kei Wong and members of the Department for their warm hospitality and many helpful conversations. The author wishes to thank Mrs. Katherine MacDougall for her excellent preparation of the text. The author is also indebted to Professor Junji Kato for his helpful criticisms of the manuscript and to Professor Shui-Nee Chow for his careful proofreading of this material.

Taro Yoshizawa

Sendai, Japan

TABLE OF CONTENTS

CHAPTER I

PRELIMINARIES

Throughout this lecture, real systems of differential equations will be considered and the following notations will be used. The real intervals $a < t < b$, $a \le t \le b$, $a \le t < b$ and $a < t \le b$ will be denoted by (a,b), $[a,b]$, $[a,b)$ and $(a,b]$, respectively. Let R denote the whole real line, i.e., $R = (-\infty,\infty)$ and I denote the interval $0 \le t < \infty$ and R^n denote Euclidean n-space. For $x \in R^n$, $|x|$ be any norm of x. For an $n \times n$ matrix $A = (a_{ij})$, define the norm $|A|$ of A by $|A| = \sup_{|x|=1} |Ax|$, where $x \in R^n$. The closure of a set S will be denoted by $\bar{S}$, and $N(\varepsilon,S)$ represents the ε-neighbrohood of S. We shall denote by $C(J \times D, R^n)$ the set of all continuous functions f defined on $J \times D$ with values in R^n, where J is a subset of R and D is a subset of R^n.

1. Liapunov Functions.

Let $f(t,x) \in C(I \times D, R^n)$, where D is an open set in R^n. For a system

$$x' = f(t,x) \qquad (' = \frac{d}{dt}), \qquad (1.1)$$

we shall consider a continuous scalar function $V(t,x)$ defined on an open set S in $R \times D$. We assume that $V(t,x)$ satisfies locally a Lipschitz condition with respect to x, that is, for each point in S there is a neighborhood U and a positive number $L(U)$ such that

$$|V(t,x)-V(t,y)| \le L(U)|x-y|$$

for any $(t,x) \in U$, $(t,y) \in U$.

Corresponding to $V(t,x)$, we define the function

$$\dot{V}_{(1.1)}(t,x) = \overline{\lim_{h\to 0^+}} \frac{1}{h}\{V(t+h,x+hf(t,x)) - V(t,x)\}. \qquad (1.2)$$

Let $x = x(t)$ be a solution of (1.1) which stays in S, and denote by $V'(t,x(t))$ the upper right-hand derivative of $V(t,x(t))$, that is,

$$V'(t,x(t)) = \overline{\lim_{h\to 0^+}} \frac{1}{h}\{V(t+h,x(t+h)) - V(t,x)\}. \qquad (1.3)$$

For a point $(t,x) \varepsilon S$ and small h, there exists a neighborhood U of (t,x) and an $L > 0$ such that $\overline{U} \subset S$, $(t+h,x+hf(t,x)) \varepsilon U$, $(t+h,x(t+h)) \varepsilon U$ and

$$|V(\tau,\xi)-V(\tau,\eta)| \le L|\xi-\eta| \quad \text{for} \quad (\tau,\xi) \varepsilon U \quad \text{and} \quad (\tau,\eta) \varepsilon U.$$

Then we have

$$V(t+h,x(t+h)) - V(t,x)$$
$$= V(t+h,x+hf(t,x)+h\varepsilon) - V(t,x) \qquad (1.4)$$
$$\le V(t+h,x+hf(t,x)) + Lh|\varepsilon| - V(t,x),$$

where ε tends to zero with h. From (1.4), it follows that

$$\overline{\lim_{h\to 0^+}} \frac{1}{h}\{V(t+h,x(t+h)) - V(t,x)\} \qquad (1.5)$$
$$\le \overline{\lim_{h\to 0^+}} \frac{1}{h}\{V(t+h,x+hf(t,x)) - V(t,x)\}.$$

On the other hand, we have

$$V(t+h,x(t+h)) - V(t,x)$$
$$\ge V(t+h,x+hf(t,x)) - Lh|\varepsilon| - V(t,x),$$

which implies that $\dot{V}_{(1.1)}(t,x) \le V'(t,x(t))$. Thus, from this and (1.5), we obtain

$$\dot{V}_{(1.1)}(t,x) = V'(t,x(t)). \qquad (1.6)$$

By the same calculation, we obtain the relation

$$\lim_{h \to 0^+} \frac{1}{h}\{V(t+h, x(t+h)) - V(t,x)\}$$

$$= \lim_{h \to 0^+} \frac{1}{h}\{V(t+h, x+hf(t,x)) - V(t,x)\}. \tag{1.7}$$

In case $V(t,x)$ has continuous partial derivatives of the first order, it is evident that

$$\dot{V}_{(1.1)}(t,x) = \frac{\partial V}{\partial t} + \frac{\partial V}{\partial x} \cdot f(t,x),$$

where "." denotes the scalar product.

Remark. In the case where $V(t,x)$ is not locally Lipschitzian with respect to x, even if the solution $x(t)$ is unique to the right, we do not necessarily have the relationship (1.6). For example, consider a function $V = \sqrt{x}$, $x \geq 0$, for the equation $x' = 2t$, $x \geq 0$. Then clearly $\dot{V}(0,0) = 0$, but for a solution $x(t) = t^2$ passing through $(0,0)$, we have $V'(0,0) = 1$.

As is known, if $\dot{V}_{(1.1)}(t,x) \leq 0$ and consequently $V'(t,x(t)) \leq 0$, the function $V(t,x(t))$ is a nonincreasing function of t, that is, $V(t,x)$ is nonincreasing along a solution of (1.1). Conversely, if $V(t,x)$ is nonincreasing along a solution of (1.1), we have $\dot{V}_{(1.1)}(t,x) \leq 0$.

The following property of the function $V(t,x)$ is important, especially in studying the behavior of solutions of perturbed systems. Let $x(s)$ and $y(s)$ be continuous and differentiable functions defined for $s \geq t$ such that $x(t) = y(t) = x$. Then, by the definition

$$V'(t,x(t)) = \overline{\lim_{h \to 0^+}} \frac{1}{h}\{V(t+h, x(t+h)) - V(t,x(t))\},$$

$$V'(t,y(t)) = \overline{\lim_{h \to 0^+}} \frac{1}{h}\{V(t+h, y(t+h)) - V(t,y(t))\}.$$

Let L be a Lipschitz constant of $V(t,x)$ in a neighborhood of the point (t,x). Then, for sufficiently small h,

$$V'(t,y(t)) \leq \varlimsup_{h \to 0^+} \frac{1}{h}\{V(t+h,x(t+h)) - V(t,y(t))\}$$

$$+ \varlimsup_{h \to 0^+} \frac{1}{h}\{V(t+h,y(t+h)) - V(t+h,x(t+h))\}$$

$$\leq \varlimsup_{h \to 0^+} \frac{1}{h}\{V(t+h,x(t+h)) - V(t,x(t))\}$$

$$+ \varlimsup_{h \to 0^+} \frac{1}{h} L|y(t+h) - x(t+h)|.$$

Thus we have

$$V'(t,y(t)) \leq V'(t,x(t)) + L|y'(t)-x'(t)|.$$

When we say that a function V(t,x) is a Liapunov function,
V(t,x) is always assumed to be a continuous scalar function which
satisfies locally a Lipschitz condition with respect to x. Consider
the system (1.1) and let V(t,x) be a Liapunov function. Suppose
that there exists a real valued continuous function $\omega(t,u)$ defined
on $0 \leq t < \infty$, $|u| < \infty$ such that for all $(t,x) \in I \times D$

$$\overset{\cdot}{V}_{(1.1)}(t,x) \leq \omega(t,V(t,x)). \tag{1.8}$$

Let $u(t,t_0,u_0)$ be the maximal solution of

$$u' = \omega(t,u), \quad u_0 = V(t_0,x_0). \tag{1.9}$$

Then, as a consequence of (1.8), a solution $x(t,t_0,x_0)$ of (1.1) and
$u(t,t_0,u_0)$ are related by the inequality

$$V(t,x(t,t_0,x_0)) \leq u(t,t_0,x_0) \tag{1.10}$$

which holds for all $t \geq t_0$ for which $x(t,t_0,x_0)$ and $u(t,t_0,x_0)$
are defined.

This is the simplest form of a very general comparison principle. The comparison principle has been widely used in dealing with a
variety of qualitative problems. It is a very important tool in

application, because it reduces the problem of determining the behavior of solutions of (1.1) to the solution of a scalar equation (1.9)
and the properties of the Liapunov function V.

The comparison principle can be verified by the following
theorem (cf. [55], [80]). Consider a scalar differential equation

$$u' = \omega(t,u), \hspace{3cm} (1.11)$$

where $\omega(t,u)$ is continuous on an open connected set $\Omega \subset R^2$.

Theorem 1.1. Let $u(t)$ be a right maximal solution of (1.11)
on an interval [a,b]. If $x(t)$ is continuous on [a,b], $x(a) \leq u(a)$
and satisfies

$$D_+ x(t) \leq \omega(t,x(t))$$

on (a,b), then $x(t) \leq u(t)$ for $a \leq t \leq b$, where $D_+ x(t) = \lim_{h \to 0^+} \frac{x(t+h)-x(t)}{h}$. Similarly, let $u(t)$ be a right minimal solution
of (1.11) on an interval [a,b]. If $x(t)$ is a continuous on [a,b],
$x(a) \geq u(a)$ and

$$D^+ x(t) \geq \omega(t,x(t))$$

on (a,b), then $x(t) \geq u(t)$ for $a \leq t \leq b$, where $D^+ x(t) = \overline{\lim}_{h \to 0^+} \frac{x(t+h)-x(t)}{h}$.

Remark. In Theorem 1.1, $h \to 0^+$ can be replaced by $h \to 0^-$.

2. Almost Periodic Functions.

Almost periodicity is a generalization of pure periodicity.
For our purpose, we shall consider an almost periodic function which
contains a parameter.

Definition 2.1. Let $f(t,x) \varepsilon C(R \times D, R^n)$, where D is an open set in R^n (more generally, a separable Banach space). $f(t,x)$ is said to be <u>almost periodic in t uniformly for x ε D</u>, if for any $\varepsilon > 0$ and any compact set S in D, there exists a positive number $\ell(\varepsilon,S)$ such that any interval of length $\ell(\varepsilon,S)$ contains a τ for which

$$|f(t+\tau,x) - f(t,x)| \leq \varepsilon \qquad\qquad (2.1)$$

for all t ε R and all x ε S.

Such a number τ in (2.1) is called an <u>ε-translation number of f(t,x)</u> and we denote by $E\{\varepsilon,f,S\}$ the set of all ε-translation numbers of f for x ε S. The following properties of translation numbers are easily verified. For a fixed compact set S,

(i) an ε-translation number is also an ε'-translation number if $\varepsilon' > \varepsilon$, and hence $E\{\varepsilon,f,S\} \subset E\{\varepsilon',f,S\}$,

(ii) if τ is an ε-translation number, so is $-\tau$,

(iii) if τ_1, τ_2 are ε_1-translation and ε_2-translation numbers, respectively, then $\tau_1 \pm \tau_2$ is an $(\varepsilon_1+\varepsilon_2)$-translation number.

<u>Definition 2.2.</u> Let $f(t,x) \in C(R \times D, R^n)$ be almost periodic in t uniformly for x ε D. Let Λ be the set of real numbers λ such that

$$\lim_{T\to\infty} \frac{1}{T} \int_0^T f(t,x) e^{-i\lambda t} dt, \quad i = \sqrt{-1} , \qquad\qquad (2.2)$$

is not identically zero for x ε D. Since D is separable, the set Λ is a countable set, say $\{\lambda_j\}$. The set consisting of all real numbers which are linear combinations of elements of the set Λ with integer coefficients is called the <u>module of f(t,x)</u>, that is, the module of $f = \{ \sum_{j=1}^{N} n_j \lambda_j; n_j, N \geq 1, \text{integer}\}$. If $\{\gamma_j\}$ is any sequence of real numbers, we say $\{\alpha_j\}$ is an <u>integral base</u> for this set, if $\{\alpha_j\}$ is linearly independent and if each γ in $\{\gamma_j\}$ is a finite linear combination of elements of $\{\alpha_j\}$ with integer co-efficients.

We shall now prove some theorems, which will be used later.

Theorem 2.1. Let $f \varepsilon C(R \times D, R^n)$ be almost periodic in t uniformly for $x \varepsilon D$. Then $f(t,x)$ is bounded and uniformly continuous on $R \times S$, S any compact set in D.

Proof. For $\varepsilon = 1$, there is an $\ell(S) > 0$ such that any interval of length $\ell(S)$ contains a τ for which

$$|f(t+\tau,x)-f(t,x)| \leq 1, \quad t \varepsilon R, \quad x \varepsilon S.$$

Let M be the maximum of $|f(t,x)|$ on $[0,\ell(S)] \times S$. It can be easily seen that for any $t \varepsilon (-\infty,\infty)$ we can find a number $\tau \varepsilon E\{1,f,S\}$ such that $t+\tau$ belongs to $[0,\ell(S)]$. Therefore $|f(t+\tau,x)| \leq M$ for $x \varepsilon S$. However,

$$|f(t+\tau,x)-f(t,x)| \leq 1, \quad x \varepsilon S,$$

and hence $|f(t,x)| \leq M+1$ for all $t \varepsilon R$ and $x \varepsilon S$.

Next we shall see the uniform continuity. For given $\varepsilon > 0$, consider an $\ell = \ell(\frac{\varepsilon}{3}, S)$ and let $\delta, 0 < \delta < 1$, be a number such that

$$|f(t_1,x)-f(t_2,x)| < \frac{\varepsilon}{3} \text{ for any } t_1, t_2, \varepsilon [0,\ell+1] \text{ and } x \varepsilon S$$

if $|t_1-t_2| < \delta$. This δ depends on ε and S, and δ exists, be-cause f is uniformly continuous on $[0,\ell+1] \times S$. Let t and t' be any two numbers such that $|t-t'| < \delta$. Then there exists a $\tau \varepsilon E\{\frac{\varepsilon}{3}, f,S\}$ such that $t+\tau \varepsilon [0,\ell+1]$ and $t'+\tau \varepsilon [0,\ell+1]$. There-fore we have

$$|f(t+\tau,x)-f(t'+\tau,x)| < \frac{\varepsilon}{3}, \quad x \varepsilon S$$

and

$$\left| f(t+\tau,x)-f(t,x) \right| < \frac{\varepsilon}{3}, \quad \left| f(t'+\tau,x)-f(t',x) \right| < \frac{\varepsilon}{3}$$

for any $t \in R$ and $x \in S$. Thus $\left| f(t,x)-f(t',x) \right| < \varepsilon$ for all t,t' such that $\left| t-t' \right| < \delta$ and for all $x \in S$. This completes the proof.

We shall now discuss the <u>normality of almost periodic functions</u>. First of all, we shall prove the following lemma.

<u>Lemma 2.1</u>. Let $f(t,x) \in C(R \times D, R^n)$ be almost periodic in t uniformly for $x \in D$, where D is an open set in R^n, and let $\{h_k\}$ be a sequence of real numbers. Then, for any $\varepsilon > 0$ and any compact set S in D, there corresponds a subsequence $\{h_{k_j}\}$ such that the norm of the difference of any pair of functions $f(t+h_{k_j},x)$, $x \in S$, is less than ε.

<u>Proof</u>. For a given $\varepsilon > 0$, there corresponds an $\ell = \ell(\frac{\varepsilon}{4},S)$ such that every interval of length ℓ contains an $\frac{\varepsilon}{4}$-translation number. For each h_k, there exists a τ_k and a γ_k such that $h_k = \tau_k+\gamma_k$, where $\tau_k \in E\{\frac{\varepsilon}{4}, f,S\}$ and $0 \le \gamma_k \le \ell$. By Theorem 2.1, $f(t,x)$ is uniformly continuous on $R \times S$, and hence, there is a $\delta(\varepsilon,S) > 0$ such that $\left| f(t',x)-f(t'',x) \right| < \frac{\varepsilon}{2}$ if $\left| t'-t'' \right| < 2\delta$ and $x \in S$. Since $0 \le \gamma_k \le \ell$, there exists a subsequence $\{\gamma_{k_j}\}$ of $\{\gamma_k\}$ such that $\gamma_{k_j} \to \gamma$ as $j \to \infty$, where γ is a limit point of the set of all γ_k and consequently $0 \le \gamma \le \ell$. Consider the h_{k_j} for which $\gamma-\delta < \gamma_{k_j} < \gamma+\delta$. Let h_{k_p}, h_{k_m} be two such values. Then we have

$$\sup_{t \in R} \left| f(t+h_{k_p},x) - f(t+h_{k_m},x) \right|$$

$$= \sup \left| f(t+\tau_{k_p}-\tau_{k_m}+\gamma_{k_p}-\gamma_{k_m},x) - f(t,x) \right|$$

$$\le \sup \left| f(t+\tau_{k_p}-\tau_{k_m}+\gamma_{k_p}-\gamma_{k_m},x) - f(t+\gamma_{k_p}-\gamma_{k_m},x) \right|$$

$$+ \sup \left| f(t+\gamma_{k_p}-\gamma_{k_m},x) - f(t,x) \right|.$$

Since $\tau_{k_p} - \tau_{k_m} \in E\{\frac{\varepsilon}{2}, f, S\}$ and $|\gamma_{k_p} - \gamma_{k_m}| < 2\delta$, we have

$$|f(t+h_{k_p}, x) - f(t+h_{k_m}, x)| < \varepsilon$$

for all $t \in R$ and $x \in S$. This proves the lemma.

Lemma 2.2. Let $f(t,x) \in C(R \times D, R^n)$ be almost periodic in t uniformly for $x \in D$ and let S be a compact set in D. Then, for any sequence $\{h_k\}$ of real numbers, there exists a subsequence $\{h_{k_j}\}$ such that the sequence of functions $\{f(t+h_{k_j}, x)\}$ is uniformly convergent on $R \times S$.

Proof. By Lemma 2.1, we can choose a subsequence $\{h_{k_j}^{(1)}\}$ such that for any two positive integers p,m

$$|f(t+h_{k_p}^{(1)}, x) - f(t+h_{k_m}^{(1)}, x)| < 1$$

for all $t \in R$ and $x \in S$. Similarly we can choose a subsequence $\{h_{k_j}^{(2)}\}$ of the sequence $\{h_{k_j}^{(1)}\}$ such that for any two positive integers p,m

$$|f(t+h_{k_p}^{(2)}, x) - f(t+h_{k_m}^{(2)}, x)| < 1/2$$

for all $t \in R$ and $x \in S$. We then choose a subsequence $\{h_{k_j}^{(3)}\}$ of $\{h_{k_j}^{(2)}\}$ such that

$$|f(t+h_{k_p}^{(3)}, x) - f(t+h_{k_m}^{(3)}, x)| < \frac{1}{3}$$

and so on. Take now the sequence of functions

$$f(t+h_{k_1}^{(1)}, x), \quad f(t+h_{k_2}^{(2)}, x), \quad f(t+h_{k_3}^{(3)}, x), \ldots.$$

Then for p,m ($p < m$), we have

$$|f(t+h_{k_p}^{(p)}, x) - f(t+h_{k_m}^{(m)}, x)| < \frac{1}{p}$$

for all $t \in R$ and $x \in S$. This shows that the sequence

$\{f(t+h_{k_j}^{(j)},x)\}$ is uniformly convergent on $R \times S$.

   Theorem 2.2. Let $f(t,x) \in C(R \times D, R^n)$ be almost periodic in
t uniformly for $x \in D$. Then, for any real sequence $\{h'_k\}$, there
exists a subsequence $\{h_k\}$ of $\{h'_k\}$ and a continuous function
$g(t,x)$ such that

$$f(t+h_k,x) \to g(t,x) \qquad\qquad (2.3)$$

uniformly on $R \times S$ as $k \to \infty$, where S is any compact set in D.
Moreover, $g(t,x)$ is also almost periodic in t uniformly for $x \in D$.

   Proof. Since D is a subset in R^n, we can prove this theorem
without difficulty. However, we shall now give a proof which can be
utilized in a more general case where D is a set in a separable
space.

   Since D is separable, there exists $\{x_\ell\}$, $x_\ell \in D$, and for
each $x \in D$ there is $\{x_{\ell_j}\}$ such that $x_{\ell_j} \to x$ and $j \to \infty$. Let
$X = \{x_\ell\}$. Since X is a countable set and one point is a compact set,
there is a subsequence $\{h_k\}$ such that $f(t+h_k,x_\ell)$ converges uniformly
for $t \in R$ and for each x_ℓ. Let S be a compact set in D and
let $\{\varepsilon_m\}$ be such that

$$\varepsilon_1 > \varepsilon_2 > \ldots > \varepsilon_m \ \ldots \to 0 \quad \text{as} \quad m \to \infty.$$

For $x \in S$, let $U(x,\delta)$ denote the δ-neighborhood of x. Then there
exists an $x^p \in X$ such that $x^p \in U(x, \frac{\varepsilon_p}{2})$. Since S is compact,
S is covered by a finite number of such neighborhoods, and hence,
there is a finite number of $x^p \in X$, say $x_1^p, x_2^p,\ldots,x_{j_p}^p$, such that
for any $x \in S$, $U(x,\varepsilon_p)$ contains a x_q^p, $q = 1,2,\ldots,j_p$. Let X^* be
the set of all points x_q^p, $p = 1,2,\ldots,q=1,2,\ldots,j_p$. Then clearly
$X^* \subset X$. Let $S^* = X^* \cup S$. Then S^* is a compact set, because for any
$\varepsilon > 0$ the number of points $x \in X^*$ such that $d(x,S) \geq \varepsilon$ is finite.

Since $f(t,x)$ is uniformly continuous on $R \times S^*$, for any $\varepsilon > 0$ there is a $\delta(\varepsilon) > 0$ such that if $|x-y| < \delta(\varepsilon)$, $x \varepsilon S^*$, $y \varepsilon S^*$, then $|f(t,x)-f(t,y)| < \varepsilon$ for $t \varepsilon R$. For any $\varepsilon > 0$ and any $x \varepsilon S$, $U(x,\delta(\varepsilon))$ contains one of x_q^p, where p is such that $\varepsilon_p < \delta(\varepsilon)$. For a fixed $x \varepsilon S$, let $x_q^p \varepsilon U(x,\delta(\varepsilon))$. Then

$$|f(t+h_k,x)-f(t+h_m,x)| \leq |f(t+h_k,x)-f(t+h_k,x_q^p)|$$

$$+ |f(t+h_k,x_q^p-f(t+h_m,x_q^p)| + |f(t+h_m,x_q^p)-f(t+h_m,x)|$$

$$< 2\varepsilon + |f(t+h_k,x_q^p)-f(t+h_m,x_q^p)|,$$

because $x \varepsilon S^*$, $x_q^p \varepsilon S^*$ and $|x-x_q^p| < \delta(\varepsilon)$. On the other hand, $f(t+h_k,x_\ell)$ is uniformly convergent on R for each x_ℓ, and hence, there exists an integer $k_0(\varepsilon,x_q^p) > 0$ such that if $m \geq k \geq k_0(\varepsilon,x_q^p)$, then $|f(t+h_k,x_q^p)-f(t+h_m,x_q^p)| < \varepsilon$. Thus, if $m \geq k \geq k_0(\varepsilon)$, where $k_0(\varepsilon) = \max\{k_0(\varepsilon,x_q^p), q = 1,2,\ldots,j_p\}$, for any $x \varepsilon S$

$$|f(t+h_k,x)-f(t+h_m,x)| < 3\varepsilon \quad \text{for all} \quad t \varepsilon R,$$

which shows that $f(t+h_k,x)$ converges uniformly on $R \times S$.

Let $g(t,x)$ be such that $f(t+h_k,x) \to g(t,x)$ as $k \to \infty$. We shall see that $g(t,x)$ is continuous on $R \times D$. Suppose that $g(t,x)$ is not continuous at (t_0,x_0). Then, for some $\varepsilon > 0$ there exist $\{\delta_m\}$, $\{t_m\}$ and $\{x_m\}$ such that $\delta_m > 0$, $\delta_m \to 0$ as $m \to \infty$, $|t_0-t_m|+|x_0-x_m| < \delta_m$ and $|g(t_0,x_0)-g(t_m,x_m)| \geq \varepsilon$. Let X be the set $\{x_1,x_2,\ldots x_m,\ldots,x_0\}$. Then X is a compact set in D. Since $f(t_0+h_k,x_0) \to g(t_0,x_0)$ and $f(t_m+h_k,x_m) \to g(t_m,x_m)$ uniformly for all m as $k \to \infty$, there exists a positive integer $k(\varepsilon,X)$ such that if $k \geq k(\varepsilon,X)$,

$$|f(t_0+h_k,x_0) - g(t_0,x_0)| < \frac{\varepsilon}{3} \qquad (2.4)$$

and

$$|f(t_m+h_k,x_m)-g(t_m,x_m)| < \tfrac{\varepsilon}{3} \quad \text{for all} \quad m. \tag{2.5}$$

Moreover, the uniform continuity of $f(t,x)$ on $R \times X$ implies that there is a $\delta(\varepsilon,X) > 0$ such that

$$|f(t,x)-f(t',y)| < \tfrac{\varepsilon}{3}$$

if $|t-t'|+|x-y| < \delta(\varepsilon,X)$, $x \in X$, $y \in X$. Therefore, if $\delta_m < \delta(\varepsilon,X)$, we have

$$|f(t_0+h_k,x_0)-f(t_m+h_k,x_m)| < \tfrac{\varepsilon}{3}. \tag{2.6}$$

Thus it follows from (2.4), (2.5) and (2.6) that $|g(t_0,x_0) -g(t_m,x_m)| < \varepsilon$ which contradicts $|g(t_0,x_0)-g(t_m,x_m)| \geq \varepsilon$. This shows that $g(t,x)$ is continuous on $R \times D$.

For any compact set $S \subset D$, take the same $\ell = \ell(\varepsilon,S)$ and τ as for $f(t,x)$. Then

$$|f(t+h_k,x)-f(t+h_k+\tau,x)| \leq \varepsilon \quad \text{for all} \quad t \in R \quad \text{and} \quad x \in S$$

implies that $|g(t,x)-g(t+\tau,x)| \leq \varepsilon$ for all $t \in R$ and $x \in S$. Thus we see that $g(t,x)$ also is almost periodic in t uniformly for $x \in D$.

Remark. For $g(t,x)$ in Theorem 2.2, we can prove that there is a sequence $\{\sigma_k\}$ such that $\sigma_k \to \infty$ as $k \to \infty$ and that $\{f(t+\sigma_k,x)\}$ converges to $g(t,x)$ uniformly on $R \times S$ for any compact subset S in D. For the proof, see [34].

Theorem 2.3. Let $f(t,x) \in C(R \times D,R^n)$ and assume that for any sequence $\{h_k\}$ of real numbers, there exists a subsequence $\{h_{k_j}\}$ of $\{h_k\}$ such that $\{f(t+h_{k_j},x)\}$ converges uniformly on $R \times S$, where S is any compact set in D. Then $f(t,x)$ is almost periodic in t uniformly for $x \in D$.

Proof. Suppose that $f(t,x)$ is not almost periodic in t uniformly for $x \in D$. Then there exists an $\varepsilon > 0$ and a compact set S in D such that for any $\ell > 0$, we can determine an interval of length ℓ which contains no ε-translation number of $f(t,x)$ for $x \in S$. Consider an arbitrary number h_1 and let (a_1,b_1) be an interval of length $> 2|h_1|$ which does not contain any ε-translation number. If we set $h_2 = \frac{1}{2}(a_1+b_1)$, then $h_2-h_1 \in (a_1,b_1)$ and consequently h_2-h_1 cannot be an ε-translation number. Define now an interval (a_2,b_2) of length $> 2(|h_1|+|h_2|)$ which does not contain any ε-translation number. Letting $h_3 = \frac{1}{2}(a_2+b_2)$, we have $h_3-h_2 \in (a_2,b_2)$ and $h_3-h_1 \in (a_2,b_2)$. Therefore h_3-h_1 and h_3-h_2 are not ε-translation numbers. In a similar way we define $h_4,h_5,\ldots$ so that none of the difference h_i-h_j is an ε-translation number. Therefore, for any i and j, $i \neq j$,

$$\sup_{\substack{t\in R \\ x\in S}} |f(t+h_i,x)-f(t+h_j,x)| = \sup_{\substack{t\in R \\ x\in S}} |f(t+h_i-h_j,x)-f(t,x)| > \varepsilon,$$

which proves that the sequence $\{f(t+h_k,x)\}$ cannot contain any uniformly convergent subsequence. This contradicts the hypothesis of the theorem. Thus $f(t,x)$ is almost periodic in t uniformly for $x \in D$.

From Theorems 2.2 and 2.3, it follows that if $f(t,x) \in C(R \times D,R^n)$ is almost periodic in t uniformly for $x \in D$, then so is every component of f, and conversely.

The following theorems can be easily proved.

Theorem 2.4. Let $f(t,x) \in C(R \times D,R)$ and $g(t,x) \in C(R \times D,R)$ be almost periodic in t uniformly for $x \in D$. Then $cf(t,x)$ (c: constant), $f^2(t,x)$, $f(t,x)+g(t,x)$ and $f(t,x)g(t,x)$ are almost periodic in t uniformly for $x \in D$. Moreover, if

$$\inf_{\substack{t \varepsilon R \\ x \varepsilon S}} |g(t,x)| = m(S) > 0, \quad S: \text{ any compact set,}$$

then $\dfrac{f(t,x)}{g(t,x)}$ is almost periodic in t uniformly for x ε D.

Remark. If we define the almost periodicity in the following way:

f(t,x) ε C(R × D,R^n) is almost periodic in t uniformly for x ε D, if for any ε > 0 there exists a positive number $\ell(\varepsilon)$ such that any interval of length $\ell(\varepsilon)$ contains a τ such that $|f(t+\tau,x)-f(t,x)| \leq \varepsilon$ for all t ε R, x ε D,

then this theorem is not true. For example, x sin t and x sin πt are periodic, and hence, they are almost periodic in t uniformly for x ε D in the above sense, where D = R. However the sum

$$x \sin t + x \sin \pi t = xf(t)$$

is not almost periodic in the above sense. Indeed, if there is an ε-translation number τ, $|xf(t+\tau)-xf(t)| \leq |x||f(t+\tau)-f(t)| \leq \varepsilon$ for x ε R implies f(t+τ) = f(t) for all t ε R, and hence, f(t) has a period τ, which is impossible.

Theorem 2.5. Let $f_k(t,x)$ ε C(R × D,R^n) be almost periodic in t uniformly for x ε D and assume that the sequence $\{f_k(t,x)\}$ converges uniformly on R × S to a function f(t,x), where S is any compact set in D. Then f(t,x) is also almost periodic in t uniformly for x ε D.

Proof. For a given ε > 0 and a compact set S, there exists a function $f_k(t,k)$ such that

$$|f(t,x)-f_k(t,x)| < \frac{\varepsilon}{3} \quad \text{for all } t \varepsilon R \text{ and } x \varepsilon S.$$

Let $\ell(\frac{\varepsilon}{3}, S)$ be an $\frac{\varepsilon}{3}$-translation number for $f_k(t,x)$. Then every interval of length $\ell(\frac{\varepsilon}{3}, S)$ contains a τ for which

$$\left| f_k(t+\tau,x) - f_k(t,x) \right| \leq \frac{\varepsilon}{3} \quad \text{for all} \quad t \in R \quad \text{and} \quad x \in S.$$

Therefore we have

$$\left| f(t+\tau,x) - f(t,x) \right| \leq \left| f(t+\tau,x) - f_k(t+\tau,x) \right| + \left| f_k(t+\tau,x) - f_k(t,x) \right|$$
$$+ \left| f_k(t,x) - f(t,x) \right| \leq \varepsilon.$$

This proves that $f(t,x)$ is almost periodic in t uniformly for $x \in D$.

Furthermore, we can characterize <u>almost periodic functions</u> in the following way.

<u>Theorem 2.6 [4]</u>. Let $f(t,x) \in C(R \times D, R^n)$ be almost periodic in t uniformly for $x \in D$. Then, for any real sequences $\{\alpha_k'\}$ and $\{\beta_k'\}$ there exist subsequences $\{\alpha_k\}$ and $\{\beta_k\}$ such that

$$\lim_{m \to \infty} \{ \lim_{\ell \to \infty} f(t+\alpha_\ell+\beta_m,x) \} = \lim_{k \to \infty} f(t+\alpha_k+\beta_k,x)$$

uniformly on $R \times S$, S: any compact set in D. The converse holds also.

<u>Proof</u>. If we set $\beta_k' = 0$ for all k, the converse follows immediately from Theorem 2.3. Now we shall prove the first part of the theorem. By Theorem 2.2, there exists a subsequence $\{\alpha_k\}$ of $\{\alpha_k'\}$ for which $f(t+\alpha_k,x) \to g(t,x)$ uniformly on $R \times S$, where S is any compact set in D, and there exists also a subsequence $\{\beta_k\}$ of $\{\beta_k'\}$ such that $g(t+\beta_k,x) \to h(t,x)$ uniformly on $R \times S$. Therefore, for any compact set S, if $k \geq k_1(\ ,S)$,

$$\left| f(t+\alpha_k,x) - g(t,x) \right| < \frac{\varepsilon}{2} \quad \text{for all} \quad t \in R \quad \text{and} \quad x \in S,$$

and hence

$$|f(t+\alpha_k+\beta_k,x)-g(t+\beta_k,x)| < \frac{\varepsilon}{2} \quad \text{for all } t \varepsilon R, \qquad (2.7)$$

$$x \varepsilon S \quad \text{and any } \beta_k.$$

Moreover, if $k \geq k_2(\varepsilon,S)$,

$$|g(t+\beta_k,x)-h(t,x)| < \frac{\varepsilon}{2} \quad \text{for all } t \varepsilon R \text{ and } x \varepsilon S. \quad (2.8)$$

Thus, if $k \geq \max\{k_1(\varepsilon,S),k_2(\varepsilon,S)\}$, it follows from (2.7) and (2.8) that

$$|f(t+\alpha_k+\beta_k,x)-h(t,x)| < \varepsilon \quad \text{for all } t \varepsilon R, x \varepsilon S,$$

which shows that $f(t+\alpha_k+\beta_k,x) \to h(t,x)$ uniformly on $R \times S$.

Theorem 2.7. Let $f(t,x) \varepsilon C(R \times D,R^n)$ be almost periodic in t uniformly for $x \varepsilon D$ and $\xi(t)$ be an almost periodic function such that $\xi(t) \varepsilon S$ for all $t \varepsilon R$, where S is a compact set in D. Then $f(t,\xi(t))$ is almost periodic in t.

Proof. Let $\{h_k'\}$ be a sequence of real numbers. Then there is a subsequence $\{h_k\}$ of $\{h_k'\}$ and almost periodic functions $g(t,x)$, $\eta(t)$ such that

$$f(t+h_k,x) \to g(t,x) \quad \text{uniformly on } R \times S,$$

$$\xi(t+h_k) \to \eta(t) \quad \text{uniformly on } R.$$

Since $g(t,x)$ is almost periodic in t uniformly for $x \varepsilon D$, $g(t,x)$ is uniformly continuous on $R \times S$, and hence, there is a $\delta(\frac{\varepsilon}{2}) > 0$ such that $|x-y| < \delta(\frac{\varepsilon}{2})$ implies

$$|g(t,x)-g(t,y)| < \frac{\varepsilon}{2}, \, x \varepsilon S, \, y \varepsilon S, \, t \varepsilon R.$$

Moreover, there exists a $k_0(\varepsilon) > 0$ such that if $k \geq k_0(\varepsilon)$,

$$\left| f(t+h_k,x)-g(t,x) \right| < \frac{\varepsilon}{2} \quad \text{for all} \quad t \ \varepsilon \ R \quad \text{and} \quad x \ \varepsilon \ S,$$

$$\left| \xi(t+h_k)-\eta(t) \right| < \delta\left(\frac{\varepsilon}{2}\right) \quad \text{for all} \quad t \ \varepsilon \ R.$$

On the other hand, for $k \geq k_0(\varepsilon)$

$$\left| f(t+h_k,\xi(t+h_k))-g(t,\eta(t)) \right|$$

$$\leq \left| f(t+h_k,\xi(t+h_k))-g(t,\xi(t+h_k)) \right| + \left| g(t,\xi(t+h_k))-g(t,\eta(t)) \right|$$

$$< \frac{\varepsilon}{2} + \frac{\varepsilon}{2} = \varepsilon,$$

because $\xi(t+h_k) \ \varepsilon \ S$ for all $t \ \varepsilon \ R$. This shows that $f(t+h_k,\xi(t+h_k))$ converges uniformly to $g(t,\eta(t))$ on R. Thus $f(t,\xi(t))$ is almost periodic in t.

We shall denote by $T(f)$ the function space consisting of all translates of f, that is, $f_\tau \ \varepsilon \ T(f)$, where

$$f_\tau(t,x) = f(t+\tau,x), \quad \tau \ \varepsilon \ R$$

Let $H(f)$ denote the uniform closure of $T(f)$ in the sense of (2.3). $H(f)$ is called the <u>hull of f</u>. By Theorem 2.2, if $f \ \varepsilon \ C(R \times D,R^n)$ is almost periodic in t uniformly for $x \ \varepsilon \ D$, so is a function in $H(f)$. By Theorem 2.2, $g \ \varepsilon \ H(f)$ can be defined by

$$\lim_{k\to\infty} f(t+h_k,x) = g(t,x) \quad \text{for some sequence} \quad \{h_k\}.$$

Here it should be noticed that even if $f(t,x)$, which is almost periodic in t uniformly for $x \ \varepsilon \ D$, satisfies locally a Lipschitz condition in x, a function in $H(f)$ does not necessarily satisfy locally a Lipschitz condition in x [64]. To see this, consider a function constructed by Sibuya

$$f(t,x) = (|x|+a(t))^{1/2},$$

where $a(t)$ is continuous on R and is almost periodic. There is a function a(t) which satisfies

>   (i) $a(t) > 0$ for all t,
>
>   (ii) $a(h_n+t) \to 0$ for $|t| \leq 1/2$ as $n \to \infty$ for some sequence
> $\{h_n\}$ with $h_n \to \infty$.

Then there exists a subsequence $\{h'_n\}$ of $\{h_n\}$ for which

$$a(h'_n+t) \to \hat{a}(t) \text{ uniformly,}$$

because a(t) is almost periodic. Therefore, $(|x|+\hat{a}(t))^{1/2}$ is in
H(f). Since $\hat{a}(t) = 0$ for $|t| \leq \frac{1}{2}$, $(|x|+\hat{a}(t))^{1/2}$ is not
Lipschitzian.

   We shall illustrate the existence of such a function a(t) by
constructing a discontinuous almost periodic function with the required properties. This function can be made continuous by one of the
standard smoothing processes. For $n = 1,2,\ldots$, let $b_n(t)$ be the
periodic function of period 2^{n+1} defined by

$$b_n(t) = \begin{cases} 0 & (0 \leq t \leq 2^n) \\ -2^{-n} & (2^n < t < 2^{n+1}). \end{cases}$$

Let $b_0(t) = 1$ for all t. Define a(t) by $a(t) = \sum\limits_{n=0}^{\infty} b_n(t)$. Since
$|b_n(t)| \leq 1/2^n$, this series is uniformly convergent, and hence a(t)
is almost periodic, because each b_n is periodic. Clearly $a(t) > 0$
for all t. If we set $\sigma_n = 2^n-1$ for $n = 1,2,\ldots$, then we can
establish by induction that

$$a(\sigma_n+t) = 2^{-n+1} \text{ on } 0 \leq t \leq 1.$$

If $h_n = 2^n - \frac{1}{2}$, then $a(h_n+t) \to 0$ for $|t| \leq \frac{1}{2}$ as $n \to \infty$.

   Theorem 2.8. Let f(t,x) and g(t,x) be almost periodic in
t uniformly for $x \in D$. If for any compact set S in D and for any

sequence of real numbers $\{\tau_k\}$ having its limit (including infinity) for which $\{f(t+\tau_k,x)\}$ is uniformly convergent on $R \times S$, $\{g(t+\tau_k,x)\}$ also is uniformly convergent on $R \times S$, then the module of $g(t,x)$ is contained in the module of $f(t,x)$.

For the proof in the case where D is a compact set, refer to [17]. Let Λ_S and Λ be the sets of λ such that

$$\Lambda_S = \{\lambda; \lim_{T\to\infty} \frac{1}{T}\int_0^T f(t,x)e^{-i\lambda t}dt \not\equiv 0 \text{ for } x \in S, \text{ S: compact set in } D\}$$

and

$$\Lambda = \{\lambda; \lim_{T\to\infty} \frac{1}{T}\int_0^T f(t,x)e^{-i\lambda t}dt \not\equiv 0 \text{ for } x \in D\}.$$

Then $\Lambda = \bigcup_S \Lambda_S$. Thus the module of f on S is contained in the module of f on D. Let $m(f,S)$ be the set of the module of f on S and let $m(f,D)$ be the set of the module of f on D. Then $m(f,D) = \bigcup_S m(f,S)$. Therefore, if for any compact set S in D, $m(f,S) \supset m(g,S)$, then clearly $m(f,D) \supset m(g,D)$. In the case where $g = g(t)$, if $m(f,S) \supset m(g)$ for some S, then $m(f,D) \supset m(g)$.

For other properties of almost periodic functions, for example, refer to [3], [15].

Appendix

We shall here show that the set Λ in Definition 2.2 is countable. Since D is separable, there exists a countable set $\{\bar{x}_k\}$, $k = 1,2,\ldots$ such that $\bar{x}_k \in D$ and that for any $x \in D$, there is a subsequence of $\{\bar{x}_k\}$ which tends to x. Consider

$$a(\lambda,x) = \lim_{T\to\infty} \frac{1}{T}\int_0^T f(t,x)e^{-i\lambda t}dt$$

for any $x \in D$. Let $\{x_k\}$ be the subset of $\{\bar{x}_k\}$ which tends to x. For each fixed x_k, the set of $\lambda_j^{(k)}$ such that $a(\lambda_j^{(k)},x_k) \neq 0$ is countable and the number of λ such that $|a(\lambda,x_k)| \geq \frac{\varepsilon}{2}$ is finite, say $\lambda_1^{(k)}, \ldots, \lambda_{j_k}^{(k)}$ (cf. [3]).

Suppose $|a(\lambda,x)| \geq \varepsilon > 0$. We have

$$\varepsilon \leq |a(\lambda,x)| \leq |a(\lambda,x)-a(\lambda,x_k)| + |a(\lambda,x_k)|.$$

Since $f(t,x)$ is almost periodic in t uniformly for $x \in D$ and the
set $\{x_1,x_2,\ldots,x_k,\ldots,x\}$ is compact, there is a $\delta(\varepsilon,x)$ such that
$|x-x_k| < \delta(\varepsilon,x)$ implies $|f(t,x)-f(t,x_k)| < \frac{\varepsilon}{2}$ for all t. Therefore,
if $|x-x_k| < \delta(\varepsilon,x)$, we have

$$|a(\lambda,x)-a(\lambda,x_k)| = |\lim_{T \to \infty} \frac{1}{T} \int_0^T \{f(t,x)-f(t,x_k)\}e^{-i\lambda t}dt|$$

$$\leq \lim_{T \to \infty} \frac{1}{T} \int_0^T |f(t,x)-f(t,x_k)|dt < \frac{\varepsilon}{2}.$$

Thus, if $|x-x_k| < \delta(\varepsilon,x)$,

$$\varepsilon \leq |a(\lambda,x_k)| + \frac{\varepsilon}{2} \quad \text{or} \quad \frac{\varepsilon}{2} \leq |a(\lambda,x_k)|.$$

Therefore λ must be $\lambda_1^{(k)}$ or .... or $\lambda_{j_k}^{(k)}$. For any $x \in D$,
there exists an $\bar{x} \in \{\bar{x}_k\}$ like x_k above. Since $\{\bar{x}_k\}$ is countable,
the set of all $\bar{x}$ is also countable, and hence the set of all λ
like $\lambda_1^{(k)}, \ldots, \lambda_{j_k}^{(k)}$ is countable, that is, the set of λ such that
$|a(\lambda,x)| \geq \varepsilon > 0$ for some $x \in D$ is countable. Setting
$\varepsilon = 1, \frac{1}{2}, \ldots, \frac{1}{m}, \ldots$, we can see that the set Λ is countable.

3. Asymptotically Almost Periodic Functions

Let $f(t)$ be a continuous vector function defined on
$I = [0,\infty)$ with values in R^n. The concept of asymptotic almost
periodicity was introduced by Fréchet [21].

Definition 3.1. $f(t)$ is said to be asymptotically almost
periodic if it is a sum of a continuous almost periodic function $p(t)$
and a continuous function $g(t)$ defined on I which tends to zero
as $t \to \infty$, that is,

$$f(t) = p(t) + q(t). \tag{3.1}$$

Theorem 3.1. Let f(t) be an asymptotically almost periodic function. Then its decomposition (3.1) is unique.

Proof. Suppose that f(t) has another decomposition f(t) = r(t) + s(t), where r(t) is almost periodic and s(t) → 0 as t → ∞ . Then we have

$$p(t) - r(t) + q(t) - s(t) = 0,$$

which implies p(t)−r(t) → 0 as t → ∞. Both p(t) and r(t) are almost periodic, and hence p(t) − r(t) is almost periodic. Since p(t)−r(t) → 0 as t → ∞, p(t)−r(t) ≡ 0. Thus we can see that the decomposition is unique.

Theorem 3.2. An asymptotically almost periodic function f(t) is bounded and is uniformly continuous on I.

Lemma 3.1. If an indefinite integral of an almost periodic function p(t) is bounded, then it is almost periodic.
For the proof, see [15].

Theorem 3.3. Suppose that an asymptotically almost periodic function f(t) is differentiable and its derivative f'(t) is also asymptotically almost periodic. Then the decomposition of f'(t) is precisely

$$f'(t) = p'(t) + q'(t), \tag{3.2}$$

where p'(t) and q'(t) are the derivatives of p(t) and q(t), respectively.

Proof. Since f'(t) is assumed to be asymptotically almost periodic, f'(t) has its decomposition

$$f'(t) = \alpha(t) + \beta(t),$$

where $\alpha(t)$ is almost periodic and $\beta(t) \to 0$ as $t \to \infty$. For any fixed h,

$$f(t+h)-f(t) = \int_t^{t+h} \alpha(s)ds + \int_t^{t+h} \beta(s)ds.$$

The first term of the right-hand side is almost periodic, since it is bounded and its derivative is almost periodic. The second term is continuous and tends to zero as $t \to \infty$. Therefore, by the uniqueness of the decomposition

$$p(t+h)-p(t) = \int_t^{t+h} \alpha(s)ds$$

and

$$q(t+h)-q(t) = \int_t^{t+h} \beta(s)ds.$$

From this it follows that $p(t)$ and $q(t)$ are differentiable and

$$p'(t) = \alpha(t) \quad \text{and} \quad q'(t) = \beta(t).$$

Now let $f(t)$ be a continuous function defined on I.

Definition 3.2. We say that $f(t)$ has the property P^*, if given $\varepsilon > 0$ there is an $\ell(\varepsilon) > 0$ and a $T(\varepsilon) \geq 0$ such that every interval of length $\ell(\varepsilon)$ contains a τ such that

$$|f(t+\tau)-f(t)| < \varepsilon \quad \text{for} \quad t \geq T(\varepsilon) \quad \text{and} \quad t+\tau > T(\varepsilon).$$

Definition 3.3. We say that $f(t)$ has the property P, if given $\varepsilon > 0$ there is an $\ell(\varepsilon) > 0$ and a $T(\varepsilon) \geq 0$ such that every interval of length $\ell(\varepsilon)$ on I contains a τ such that

$$|f(t+\tau)-f(t)| < \varepsilon \quad \text{for} \quad t \geq T(\varepsilon).$$

Lemma 3.2. The property P^* is equivalent to the property P.

Proof. Evidently, the property P^* implies the property P.

Suppose now that f(t) has the property P. Consider a nonpositive segment L of length $\ell(\varepsilon)$. In the case where L contains the origin, take $\tau = 0$. In other cases, take L* which is symmetric to L with respect to the origin. Then, for some $\tau^* \varepsilon L^*$,

$$|f(t+\tau^*)-f(t)| < \varepsilon \quad \text{for} \quad t \geq T(\varepsilon).$$

If we set $\tau = -\tau^*$, then $\tau \varepsilon L$. Set $\sigma = t+\tau$. Then $t = \sigma+\tau^*$. Since $\tau^* \varepsilon L^*$ and f(t) has the property P,

$$|f(\sigma+\tau^*)-f(\sigma)| < \varepsilon \quad \text{for} \quad \sigma \geq T(\varepsilon),$$

which implies that

$$|f(t)-f(t+\tau)| < \varepsilon \quad \text{for} \quad t+\tau \geq T(\varepsilon) \quad \text{and} \quad t \geq T(\varepsilon).$$

This shows that the property P implies the property P*.

Definition 3.4. We say that f(t) has the property L, if for any sequence $\{h_k\}$ such that $h_k > 0$ and $h_k \to \infty$ as $k \to \infty$, we can select a subsequence $\{h_{k_j}\}$ such that $f(t+h_{k_j})$ converges uniformly on I.

Theorem 3.4. If f(t), $t \varepsilon I$, is asymptotically almost periodic, f(t) has the property P.

Proof. Since f(t) is asymptotically almost periodic, it has the decomposition f(t) = p(t)+q(t), where p(t) is almost periodic and $q(t) \to 0$ as $t \to \infty$. Since p(t) is almost periodic, for given $\varepsilon > 0$ there is an $\ell(\varepsilon) > 0$ such that every interval of length $\ell(\varepsilon)$ on I contains τ such that $|p(t+\tau)-p(t)| < \frac{\varepsilon}{3}$. Moreover, there is a $T(\varepsilon) \geq 0$ such that $|q(t)| < \frac{\varepsilon}{3}$ for $t \geq T(\varepsilon)$. Thus we have

$$|f(t+\tau)-f(t)| \leq |p(t+\tau)-p(t)| + |q(t+\tau)| + |q(t)|,$$

and hence $|f(t+\tau)-f(t)| < \varepsilon$ for $t \geq T(\varepsilon)$. This shows that $f(t)$ has the property P.

Theorem 3.5. If $f(t)$, $t \varepsilon I$, is asymptotically almost periodic, $f(t)$ has the property L.

Proof. For any sequence $\{h_k\}$ such that $h_k > 0$ and $h_k \to \infty$ as $k \to \infty$, we have

$$f(t+h_k) = p(t+h_k) + q(t+h_k).$$

Since $q(t) \to 0$ as $t \to \infty$, there is a positive integer $k_0(\varepsilon)$ such that if $m,k \geq k_0(\varepsilon)$, then

$$|q(t+h_k)-q(t+h_m)| < \varepsilon \text{ for all } t \varepsilon I.$$

Moreover, there exists a subsequence $\{h_{k_j}\}$ of $\{h_k\}$ and a positive integer $j_0(\varepsilon)$ such that if $j,\ell \geq j_0(\varepsilon)$, then $|p(t+h_{k_j}-p(t+h_{k_\ell})| < \varepsilon$ for all $t \varepsilon R$, because $p(t)$ is almost periodic. Thus there is a positive integer $N(\varepsilon)$ such that if $j,\ell \geq N(\varepsilon)$,

$$|p(t+h_{k_j})-p(t+h_{k_\ell})| < \varepsilon, \quad |q(t+h_{k_j})-q(t+h_{k_\ell})| < \varepsilon$$

for all $t \varepsilon I$. This implies that if $j,\ell \geq N(\varepsilon)$ and $t \varepsilon I$, we have $|f(t+h_{k_j})-f(t+h_{k_\ell})| < 2\varepsilon$. This shows that $f(t)$ has the property L.

Theorem 3.6. If $f(t)$, $t \varepsilon I$, has the property P, then $f(t)$ is bounded and uniformly continuous on I.

Proof. For given $\varepsilon > 0$, there is an $\ell(\varepsilon) > 0$ and a $T(\varepsilon) \geq 0$ such that every interval of length $\ell(\varepsilon)$ on I contains a τ such that

$$|f(t+\tau)-f(t)| < \varepsilon \text{ for } t \geq T(\varepsilon).$$

Let $t' > T(\varepsilon)+\ell(\varepsilon)$. Then $[t'-T-\ell,t'-T]$ is an interval of length

$\ell(\varepsilon)$ on I. Hence there is a $\tau \in [t'-T-\ell, t'-T]$ such that $|f(t+\tau)-f(t)| < \varepsilon$ for $t \geq T(\varepsilon)$. If we set $t = t'-\tau$, then t $T \leq t \leq T+\ell$. Thus we have $|f(t')-f(t)| < \varepsilon$ or $|f(t')| \leq |f(t)|+\varepsilon \leq M_1+\varepsilon$, where $M_1 = \max\{|f(t)| : T \leq t \leq T+\ell\}$. For $t' \in [0, T+\ell]$, there is a constant M_2 such that $M_2 = \max\{|f(t')| : 0 \leq t' \leq T+\ell\}$. Therefore $|f(t)|$ is bounded on I.

The uniform continuity of $f(t)$ can be proved by the same argument as in the proof of Theorem 2.1.

Theorem 3.7. If $f(t)$, $t \in I$, has the property P, then $f(t)$ has the property L.

Proof. Let $\{h_k\}$ be a sequence such that $h_k > 0$ and $h_k \to \infty$ as $k \to \infty$. For a fixed β, $-\infty < \beta \leq 0$, if k is sufficiently large, say $k > K_1$, $f(t+h_k)$ is defined on $\beta \leq t < \infty$ for all k. By Theorem 3.6, $f(t)$ is bounded and is uniformly continuous for $t \geq 0$ and hence $\{f(t+h_k)\}$ is uniformly bounded and is equicontinuous for $t \geq \beta$. Therefore there is a subsequence $\{f(t+h_k')\}$ of $\{f(t+h_k)\}$ which converges to a continuous function $p(t)$ defined on $(-\infty, \infty)$ uniformly on any compact interval in $(-\infty, \infty)$. By the property P, for given $\varepsilon > 0$ there exists an $\ell = \ell(\varepsilon) > 0$ and a $T(\varepsilon) \geq 0$ and a $\tau_k \in [h_k'-\ell, h_k']$ such that

$$|f(t+\tau_k)-f(t)| < \varepsilon \quad \text{for} \quad t \geq T(\varepsilon), \qquad (3.3)$$

where τ_k is positive if k is sufficiently large, say $k > K_2$. Let $\ell_k = h_k'-\tau_k$. Then $0 \leq \ell_k \leq \ell$. By (3.3), changing t into $t+\ell_k$,

$$|f(t+h_k')-f(t+\ell_k)| < \varepsilon \quad \text{for} \quad t+\ell_k \geq T(\varepsilon). \qquad (3.4)$$

Since $0 \leq \ell_k \leq \ell$, there exists a subsequence such that

$$\lim_{j \to \infty} \ell_{k_j} = \ell^*, \quad 0 \le \ell^* \le \ell.$$

Consider $f(t+h'_{k_j})$ on $[0,\infty)$. If k_j is sufficiently large, by (3.4)

$$\left| f(t+h'_{k_j}) - f(t+\ell_{k_j}) \right| < \varepsilon \quad \text{for} \quad t \ge T(\varepsilon),$$

where $t \ge T(\varepsilon)$ and $\ell_k \ge 0$ imply $t+\ell_k \ge T(\varepsilon)$. By Theorem 3.6,
$f(t)$ is uniformly continuous for $t \ge 0$, and hence there is an integer
$j_0(\varepsilon) > 0$ such that $j \ge j_0(\varepsilon)$ implies

$$\left| f(t+\ell_{k_j}) - f(t+\ell^*) \right| < \varepsilon \quad \text{for} \quad t \ge 0.$$

Thus, if $j \ge j_0(\varepsilon)$ and $t \ge T(\varepsilon)$, we have

$$\left| f(t+h'_{k_j}) - f(t+\ell^*) \right| < 2\varepsilon. \qquad (3.5)$$

However, for any t , $f(t+h'_{k_j}) \to p(t)$ as $j \to \infty$, and therefore, by
(3.5)

$$\left| p(t) - f(t+\ell^*) \right| \le 2\varepsilon \quad \text{for} \quad t \ge T(\varepsilon).$$

Therefore $\left| f(t+h'_{k_j}) - p(t) \right| < 4\varepsilon$ for $j \ge j_0(\varepsilon)$ and $t \ge T(\varepsilon)$.

On the other hand, for t such that $0 \le t < T(\varepsilon)$, there is
an integer $j'_0(\varepsilon) > 0$ such that if $j \ge j'_0(\varepsilon)$ and $0 \le t \le T(\varepsilon)$,
then $\left| f(t+h'_{k_j}) - p(t) \right| < 4\varepsilon$. Thus, if $j \ge j_0(\varepsilon) + j'_0(\varepsilon)$ and $t \ge 0$,

$$\left| f(t+h'_{k_j}) - p(t) \right| < 4\varepsilon.$$

Clearly $j_0(\varepsilon)$ and $j'_0(\varepsilon)$ depend only on ε . This completes the
proof.

Theorem 3.8. If $f(t)$, $t \in I$, has the property P, then the
function $p(t)$ in the proof of Theorem 3.7 is an almost periodic
function.

Proof. By Lemma 3.2, $f(t)$ has the property P*, that is, for any $\varepsilon > 0$ there is an $\ell(\varepsilon) > 0$ and a $T(\varepsilon) \geq 0$ such that every interval of length $\ell(\varepsilon)$ contains a τ such that

$$|f(t+\tau)-f(t)| < \varepsilon \quad \text{for} \quad t \geq T(\varepsilon) \quad \text{and} \quad t+\tau \geq T(\varepsilon).$$

Therefore we have $|f(t+\tau+h_{k_j})-f(t+h_{k_j})| < \varepsilon$ for $t \geq T(\varepsilon)-h_{k_j}$ and $t+\tau \geq T(\varepsilon)-h_{k_j}$. For a fixed $t \varepsilon (-\infty,\infty)$, $t, t+\tau \geq T(\varepsilon)-h_{k_j}$ if j is sufficiently large. Letting $j \to \infty$, we have

$$|p(t+\tau)-p(t)| \leq \varepsilon \quad \text{for all} \quad t \varepsilon (-\infty,\infty).$$

This shows that $p(t)$ is almost periodic.

   Theorem 3.9. If $f(t)$, $t \varepsilon I$, has the property P, then $f(t)$ is asymptotically almost periodic.

   Proof. For $\varepsilon_k > 0$, there exists an $\ell_k > 0$ and $T_k \geq 0$ such that every interval $[k,k+\ell_k]$ contains a τ_k such that $|f(t+\tau_k)-f(t)| < \varepsilon_k$ for $t \geq T_k$. Since $\tau_k \geq k$, $\tau_k \to \infty$ as $k \to \infty$. By Theorems 3.7 and 3.8, $f(t+\tau_k)$ has a subsequence $f(t+\tau_{k_j})$ which converges to an almost periodic function $p(t)$ uniformly on the interval $[0,\infty)$. Assume that $\varepsilon_k \to 0$ as $k \to 0$. Let η_{k_j} be defined by

$$\eta_{k_j} = \sup_{0 \leq t < \infty} |f(t+\tau_{k_j})-p(t)|.$$

Then $\eta_{k_j} \to 0$ as $j \to \infty$. Set $q(t) = f(t)-p(t)$ for $t \geq 0$. Then we have

$$|q(t)| \leq |f(t)-f(t+\tau_{k_j})| + |f(t+\tau_{k_j})-p(t)| < \varepsilon_{k_j}+\eta_{k_j}$$

for $t \geq T_{k_j}$. This shows that $q(t) \to 0$ as $t \to \infty$. Thus $f(t)$ is asymptotically almost periodic.

   Theorem 3.10. If $f(t)$, $t \varepsilon I$, has the property L, then

$f(t)$ has the property P.

Proof. Suppose $f(t)$ does not have the property P. Then there exists some $\varepsilon > 0$ and for any ℓ and any T, there is an $a(\ell,T) > 0$ such that for any $\tau \in [a,a+\ell]$, there is a $t(\ell,T,a,\tau) \geq T$ such that $|f(t+\tau)-f(t)| \geq \varepsilon$. For an integer $k > 0$, denote by L_k the interval $[a_k,a_k+k]$ corresponding to $a_k = a(k,k)$. Take an $h_1 \in L_{k_1}$. The length of L_{k_1} is k_1 and $h_1 \geq 0$. For a $k_2 > h_1$, $k_2 > k_1$, set $h_2 = a_{k_2} + k_2 + h_1$. Then $h_2 > h_1$ and $h_2 > k_2$ and $h_2 - h_1 = a_{k_2} + k_2 \in L_{k_2}$. Now we assume that there are $h_1,h_2,\ldots,h_{s-1}$ and $k_1,k_2,\ldots,k_{s-1}$ such that

$$0 \leq h_1 < h_2 < \ldots < h_{s-1},$$
$$k_1 < k_2 < \ldots < k_{s-1},$$
$$h_q > k_q > h_{q-1} \qquad (q = 2,\ldots,s-1)$$

and that $h_{s-1}-h_p \in L_{k_{s-1}}$ for $p = 1,2,\ldots,s-2$, that is,

$$0 \leq a_{k_{s-1}} \leq h_{s-1}-h_{s-2} \leq h_{s-1}-h_{s-3} \leq \ldots \leq h_{s-1}-h_1 < a_{k_{s-1}}+k_{s-1}.$$

Now take h_s such that $h_s = a_{k_s} + k_s + h_1$, where $k_s > k_{s-1}$ and $k_s > h_{s-1}$. Then

$$h_s > k_s > h_{s-1}.$$

For $p = 1,2,\ldots,s-1$, we have

$$h_s-h_p = a_{k_s}+k_s+h_1-h_p \leq a_{k_s}+k_s,$$

$$h_s-h_p = a_{k_s}+k_s-h_p+h_1 \geq a_{k_s},$$

because $h_1-h_p \leq 0$ and $k_s > h_p$. This shows that $h_s-h_p \in L_{k_s}$ for $p = 1,2,\ldots,s-1$. Thus we have a sequence

$$0 \leq h_1 < h_2 < \ldots < h_s < \ldots, k_1 < k_2 < \ldots < k_s < \ldots.$$

and

$$L_1', L_2', \ldots, L_s', \ldots,$$

where $L_s' = L_{k_s}$ and $h_s < k_{s+1} < h_{s+1}$. Moreover, $h_s \geq k_s$ implies that $h_s \to \infty$ as $s \to \infty$.

Since $f(t)$ has the property L, there exists a subsequence $\{h_{s_m}\}$ of $\{h_s\}$ for which $f(t+h_{s_m})$ converges uniformly on I. For a fixed m,

$$h_{s_m} - h_{s_1}, h_{s_m} - h_{s_2}, \ldots, h_{s_m} - h_{s_{m-1}} \quad \varepsilon \ L_{s_m}'.$$

For $\varepsilon > 0$, there is an integer $M(\varepsilon) > 0$ such that for any $t \geq 0$

$$\left| f(t+h_{s_m}) - f(t+h_{s_{m+1}}) \right| < \varepsilon \quad \text{if} \quad m \geq M(\varepsilon).$$

For an $m \geq M(\varepsilon)$, set $\sigma = t+h_{s_m}$. Then we have

$$f(\sigma) - f(\sigma+h_{s_{m+1}} - h_{s_m}) \Big| < \varepsilon \quad \text{for all} \quad \sigma \geq h_{s_m}.$$

If we set $\tau = h_{s_{m+1}} - h_{s_m}$, then $\tau \varepsilon L_{s_{m+1}}'$, and moreover $k_{s_{m+1}} > h_{s_m}$. Thus

$$\left| f(t) - f(t+\tau) \right| < \varepsilon \quad \text{for all} \quad t \geq h_{s_m}$$

or

$$\left| f(t) - f(t+\tau) \right| < \varepsilon \quad \text{for all} \quad t \geq k_{s_{m+1}}.$$

This contradicts that for $\ell = T = k_{s_{m+1}}$, there is a $t \geq k_{s_{m+1}}$ such that $\left| f(t+\tau) - f(t) \right| \geq \varepsilon$ for any $\tau \varepsilon [a_{k_{s_{m+1}}}, a_{k_{s_{m+1}}} + k_{s_{m+1}}]$. Thus $f(t)$ has the property P.

Thus we can see that the following three properties are equivalent;

    (i)   <u>$f(t)$ is asymptotically almost periodic,</u>

    (ii)   <u>$f(t)$ has the property P</u>

and

(iii) f(t) has the property L.

4. Quasi-Periodic Functions

Let $f(t,x)$ be a continuous function defined on $R \times D$ with values in R^n, where D is an open set in R^n. We denote by e_j a unit vector in R^k such that the j-th component is 1 and the others are zero. Let e be a vector in R^k such that all of the components are 1.

Definition 4.1. The function $f(t,x)$ is said to be quasi-periodic in t, if there is a finite number of nonzero real numbers $\omega_1, \omega_2, \ldots, \omega_k$ and a function $F(u,x)$, where $F(u,x) \in C(R^k \times D, R^n)$ such that

$$F(u+\omega_j e_j,x) = F(u,x) \quad \text{for all} \quad u \in R^k \quad \text{and} \quad x \in D, \ j = 1,2,\ldots,k$$

and

$$F(t \, e,x) = f(t,x) \quad \text{for all} \quad t \in R \quad \text{and} \quad x \in D.$$

Remark. Without loss of generality, we can assume that $\omega_j > 0$ $(j = 1,2,\ldots,k)$ and $\{\frac{2\pi}{\omega_1}, \ldots, \frac{2\pi}{\omega_k}\}$ is linearly independent.

Theorem 4.1. Let $f(t,x) \in C(R \times D, R^n)$, where D is an open set in R^n. The function $f(t,x)$ is quasi-periodic in t if and only if it is almost periodic in t uniformly for $x \in D$ and its module has a finite integral base. Namely, the quasi-periodic function f is an almost periodic function with Fourier series

$$f(t,x) \sim \sum_m a_m(x) \exp\{2\pi i t (\frac{m_1}{\omega_1} + \ldots + \frac{m_k}{\omega_k})\}, \quad i = \sqrt{-1},$$

where $\omega_1, \ldots, \omega_k$ are some real numbers and $m = (m_1, \ldots, m_k)$ for integers $m_1, m_2, \ldots, m_k$.

Proof. Let $f(t,x) \in C(R \times D, R^n)$ be quasi-periodic in t. By the definition, there is a finite number of real numbers

$\omega_1,\ldots,\omega_k$ and a function $F(u,x)$ such that $F(te,x) = f(t,x)$ and $F(u+\omega_j e_j,x) = F(u,x)$, $j = 1,2,\ldots,k$. To show that $f(t,x)$ is almost periodic, it is sufficient to see that for any sequence $\{\tau_p\}$, there exists a subsequence $\{\tau_{p_j}\}$ such that $\{f(t+\tau_{p_j},x)\}$ converges uniformly on $R \times S$ for any compact set S in D. τ_p can be written as

$$\tau_p = s_\ell^p + n_\ell^p \omega_\ell, \qquad \ell = 1,2,\ldots,k$$

for $s_\ell^p \varepsilon [0,\omega_\ell]$ and integers n_ℓ. Therefore there is a subsequence $\{\tau_{p_j}\}$ of $\{\tau_p\}$ such that $s_\ell^{p_j}$ tends to some s_ℓ in $[0,\omega_\ell]$ as $j \to \infty$. Since we have

$$|f(t+\tau_{p_j},x) - F(t+s_1,\ldots,t+s_k,x)|$$

$$= |F(t+s_1^{p_j},\ldots,t+s_k^{p_j},x) - F(t+s_1,\ldots,t+s_k,x)|$$

and $F(u,x)$ is continuous on $R^k \times D$, we can see that $\{f(t+\tau_{p_j},x)\}$ converges uniformly on $R \times S$ for any compact set S in D.

Next we shall show that the module of $f(t,x)$ has an integral base $\{\frac{2\pi}{\omega_1},\ldots,\frac{2\pi}{\omega_k}\}$. To see this, we shall prove that if

$$\lim_{T\to\infty} \frac{1}{T} \int_0^T f(t,x)e^{-i\lambda t}dt = M(f(t,x)e^{-i\lambda t}) \neq 0$$

for some $x \varepsilon D$, then $\lambda = 2\pi(\frac{m_1}{\omega_1} +\ldots+ \frac{m_k}{\omega_k})$ for some integers $m_1,\ldots,m_k$. Let λ and x be such that $M(f(t,x)e^{-i\lambda t}) \neq 0$. It is known [32] that there is a function $F_\varepsilon(u,x)$ such that $F_\varepsilon(u,x) \to F(u,x)$ uniformly on R^k as $\varepsilon \to 0$. $F_\varepsilon(u,x)$ is given by

$$F_\varepsilon(u,x) = \sum_m \hat{F}_m(x)\exp\{i2\pi <\frac{m}{\omega},u> -2\pi|\frac{m}{\omega}|\varepsilon\} \qquad (4.1)$$

for $\varepsilon > 0$ where $m = (m_1,\ldots,m_k)$ for integers

$$m_1,\ldots,m_k, \; \frac{m}{\omega} = (\frac{m_1}{\omega_1},\ldots,\frac{m_k}{\omega_k}), <\frac{m}{\omega},u> = \frac{m_1}{\omega_1}u_1 +\ldots+ \frac{m_k}{\omega_k}u_k$$

$$\text{and } |\frac{m}{\omega}| = <\frac{m}{\omega},\frac{m}{\omega}>^{1/2}.$$

By the absolute convergence of (4.1), for any $\eta > 0$ there is an integer $N = N(\eta,\varepsilon) > 0$ such that

$$\left| F_\varepsilon(u,x) - \sum_{|m| \leq N} \hat{F}_m(x)\exp\{i2\pi <\tfrac{m}{\omega},u> - 2\pi|\tfrac{m}{\omega}|\varepsilon\} \right| < \eta \quad \text{on} \quad R^k.$$

Therefore, for sufficiently small $\varepsilon > 0$,

$$\left| F(u,x) - \sum_{|m| \leq N} \hat{F}_m(x)\exp\{i2\pi <\tfrac{m}{\omega},u> - 2\pi|\tfrac{m}{\omega}|\varepsilon\} \right| < 2\eta \quad \text{on} \quad R^k.$$

Let $u = te$ and consider the mean value. Then we have

$$\left| M(f(t,x)e^{-i\lambda t}) - \sum_{|m| \leq N} \hat{F}_m(x)\exp\{-2\pi|\tfrac{m}{\omega}|\varepsilon\} \right.$$

$$\left. \times M(\exp\{i(\tfrac{m_1}{\omega_1}2\pi + \ldots + \tfrac{m_k}{\omega_k}2\pi - \lambda\}t) \right| \leq 2\eta. \tag{4.2}$$

Since $M(f(t,x)e^{-i\lambda t}) \neq 0$ and η is arbitrary, we have

$$M(\exp\{i(\tfrac{m_1}{\omega_1}2\pi + \ldots + \tfrac{m_k}{\omega_k}2\pi - \lambda\}t) \neq 0$$

for some integers $m_1,\ldots,m_k$. Thus we have

$$\lambda = 2\pi(\tfrac{m_1}{\omega_1} + \ldots + \tfrac{m_k}{\omega_k}).$$

Now let $f(t,x)$ be almost periodic in t uniformly for $x \in D$ and assume that the module of $f(t,x)$ has an integral base $\{\tfrac{2\pi}{\omega_1},\ldots,\tfrac{2\pi}{\omega_k}\}$, where $\omega_1,\ldots,\omega_k$ are some real numbers. For any compact set S in D and any $\varepsilon > 0$, there exists a trigonometric polynomial $P(t,x;S,\varepsilon)$ such that

$$\left| f(t,x) - P(t,x;S,\varepsilon) \right| < \varepsilon \quad \text{on} \quad R \times S,$$

which is defined by

$$P(t,x;S,\varepsilon) = \sum_{q=1}^{q(\varepsilon)} a_q(x;S,\varepsilon)\exp\{2\pi it(\tfrac{m_1^q}{\omega_1} + \ldots + \tfrac{m_k^q}{\omega_k})\},$$

where $q(\varepsilon)$, $m_1^q,\ldots,m_k^q$ are integers and $a_q(x;S,\varepsilon)$ is continuous in x ε S. For the details, see [15, pp. 152-155].

Define $F(u,x;S,\varepsilon)$ by

$$F(u,x;S,\varepsilon) = \sum_{q=1}^{q(\varepsilon)} a_q(x;S,\varepsilon) \exp\{2\pi i(\frac{m_1^q}{\omega_1} u_1 + \ldots + \frac{m_k^q}{\omega_k} u_k)\}.$$

Then it is continuous on $R \times S$, is periodic in u_j with period ω_j ,
$j = 1,\ldots,k$, and satisfies

$$F(te,x;S,\varepsilon) = P(t,x;S,\varepsilon).$$

Therefore we have

$$|f(t,x)-F(te,x;S,\varepsilon)| < \varepsilon \quad \text{on} \quad R \times S. \qquad (4.3)$$

However, the set of all values of $F(te,x;S,\varepsilon)$ for $t \in R$ is
everywhere dense in the set of all values of $F(u,x;S,\varepsilon)$ for $u \in R^k$
(cf. pp. 35-37 in [3]). Therefore (4.3) implies that for any $\varepsilon > 0$
and any $\eta > 0$,

$$|F(u,x;S,\varepsilon)-F(u,x;S,\eta)| < \varepsilon+\eta \quad \text{on} \quad R^k \times S.$$

This shows that $\{F(u,x;S,\varepsilon)\}$ converges uniformly on $R^k \times S$ as
$\varepsilon \to 0$. Let $F(u,x;S)$ be the limit function. Then it is continuous
on $R^k \times S$, is periodic in u_j with period ω_j and $F(te,x;S) =$
$f(t,x)$ on $R \times S$.

Let $F(u,x;S_1)$ and $F(u,x;S_2)$ be the functions for S_1 and
S_2 , where S_1, S_2 are compact. Since $F(te,x;S_1) = F(te,x;S_2)$ for
$x \in S_1 \cap S_2$, we have

$$F(u,x;S_1) = F(u,x;S_2) \quad \text{on} \quad R^k \times (S_1 \cap S_2). \qquad (4.4)$$

Since the one point set $\{x\}$ in D is compact, there corresponds a
function $F(u,x;\{x\})$ which is continuous on $R^k \times \{x\}$, is periodic in
u_j with period ω_j and $F(te,x;\{x\}) = f(t,x)$ for $t \in R$. Consider
a function $G(u,x)$ on $R^k \times D$ defined by

$$G(u,x) = F(u,x;\{x\}). \tag{4.5}$$

Then we have

$$G(u+\omega_j e_j,x) = F(u+\omega_j e_j,x;\{x\}) = F(u,x;\{x\}) = G(u,x) \tag{4.6}$$

for $(u,x) \in R^k \times D$ and $j = 1,2,\ldots,k$, and also

$$G(te,x) = F(te,x;\{x\}) = f(t,x) \quad \text{for} \quad (t,x) \in R \times D. \tag{4.7}$$

Now we shall show the continuity of $G(u,x)$, that is, if for $(u_0,x_0) \in R^k \times D$ and $(u_p,x_p) \in R^k \times D$ and $(u_p,x_p) \to (u_0,x_0)$ as $p \to \infty$, then $G(u_p,x_p) \to G(u_0,x_0)$ as $p \to \infty$. Let S be the set $\{x_0,x_1,\ldots,x_p,\ldots\}$. Then S is compact. By (4.4),

$$F(u,x;S) = F(u,x;\{x\}) \quad \text{for} \quad (u,x) \in R^k \times S,$$

and hence $F(u,x;S) = G(u,x)$ for $(u,x) \in R^k \times S$. By the continuity of $F(u,x;S)$,

$$\lim_{p\to\infty} G(u_p,x_p) = \lim_{p\to\infty} F(u_p,x_p;S) = F(u_0,x_0;S) = G(u_0,x_0).$$

This completes the proof. This proof is due to [53].

5. Boundary Value Problem

In this section we shall discuss the two point boundary value problem for an equation of the second order

$$x'' = f(t,x,x'). \tag{5.1}$$

Lemma 5.1. Suppose that $f(t,x,y)$ is continuous on $a \leq t \leq b$, $|x| < \infty$, $|y| < \infty$ and $|f(t,x,y)| \leq L$ for some constant $L > 0$. Then, for any pair of constants A,B, there exists a solution $x(t)$ of (5.1) which satisfies the conditions

$$x(a) = A, \quad x(b) = B. \tag{5.2}$$

Let D now be a domain such that $a \leq t \leq b$, $\alpha(t) \leq x \leq \beta(t)$, where $\alpha(t)$ and $\beta(t)$ are twice differentiable on $a \leq t \leq b$ and $\alpha(t) \leq \beta(t)$. We assume that $f(t,x,y)$ is defined and continuous on $D \times R$ and that

$$\alpha''(t) \geq f(t,\alpha(t),\alpha'(t)) \tag{5.3}$$

and

$$\beta''(t) \leq f(t,\beta(t),\beta'(t)). \tag{5.4}$$

For a sufficiently large $M > 0$, define $g(t,x,y)$ by

$$g(t,x,y) = \begin{cases} f(t,x,M) & (y > M) \\ f(t,x,y) & (|y| \leq M) \\ f(t,x,-M) & (y < -M) \end{cases}$$

and define $f^*(t,x,y)$ by

$$f^*(t,x,y) = \begin{cases} g(t,\beta(t),y) + \dfrac{x-\beta(t)}{x-\beta(t)+1} & (x > \beta(t)) \\ g(t,x,y) & (\alpha(t) \leq x \leq \beta(t)) \\ g(t,\alpha(t),y) - \dfrac{\alpha(t)-x}{\alpha(t)-x+1} & (x < \alpha(t)). \end{cases}$$

Then $f^*(t,x,y)$ is continuous and bounded on $a \leq t \leq b$, $|x| < \infty$, $|y| < \infty$, and clearly

$$f^*(t,x,y) = f(t,x,y)$$

on $a \leq t \leq b$, $\alpha(t) \leq x \leq \beta(t)$, $-M \leq y \leq M$. By Lemma 5.1, the equation

$$x'' = f^*(t,x,x') \tag{5.5}$$

has a solution $x(t)$ which satisfies the condition (5.2). Let $K > 0$ be a constant such that

$$|\alpha'(t)| < K \quad |\beta'(t)| < K \quad \text{for } t \ \varepsilon \ [a,b]. \tag{5.6}$$

If $M \geq K$, the solution $x(t)$ of (5.5) such that $\alpha(a) \leq x(a) \leq \beta(a)$,

$\alpha(b) \le x(b) \le \beta(b)$ satisfies

$$\alpha(t) \le x(t) \le \beta(t) \text{ on } a \le t \le b. \qquad (5.7)$$

This will be proved in the following way.

Suppose $x(t) > \beta(t)$ at some t. Then there is a ξ, $a < \xi < b$, such that

$$x(\xi) > \beta(\xi), \ x'(\xi) = \beta'(\xi) \text{ and } x''(\xi) \le \beta''(\xi),$$

because $x(a) \le \beta(a)$ and $x(b) \le \beta(b)$. Therefore we have

$$x''(\xi) = f^*(\xi, x(\xi), x'(\xi)) > g(\xi, \beta(\xi), x'(\xi)).$$

However, $x'(\xi) = \beta'(\xi)$ and $|\beta'(\xi)| < K \le M$, and hence $g(\xi, \beta(\xi), x'(\xi)) = f(\xi, \beta(\xi), \beta'(\xi))$. Thus $x''(\xi) > f(\xi, \beta(\xi), \beta'(\xi))$. By (5.4), $x''(\xi) > \beta''(\xi)$, which contradicts $x''(\xi) \le \beta''(\xi)$. Therefore $x(t) \le \beta(t)$ on $a \le t \le b$. The same argument shows that $x(t) \ge \alpha(t)$ on $a \le t \le b$.

Thus, if we see that $|x'(t)| \le M$ for $t \in [a,b]$, this solution $x(t)$ is a solution of (5.1). To see this, we shall apply Liapunov's second method.

Theorem 5.1. Let $\phi(t)$ be a function defined on $a \le t \le b$ such that $\alpha(t) \le \phi(t) \le \beta(t)$ on $a \le t \le b$ and

$$|\phi(t) - \phi(s)| \le L|t-s|, \quad a \le t, \ s \le b,$$

where $L > 0$ is a constant. We assume that there exist four Liapunov functions $V_i(t,x,y)$ and $W_i(t,x,y)$, $i = 1,2$, such that $V_1(t,x,y)$ is defined on $a \le t \le b$, $\alpha(t) \le x \le \phi(t)$, $y \ge K$, where $K > 0$ can be large, $V_2(t,x,y)$ is defined on $a \le t \le b$, $\phi(t) \le x \le \beta(t)$, $y \le -K$, $W_1(t,x,y)$ is defined on $a \le t \le b$, $\phi(t) \le x \le \beta(t)$, $y \ge K$ and $W_2(t,x,y)$ is defined on $a \le t \le b$, $\alpha(t) \le x \le \phi(t)$, $y \le -K$.

These functions are assumed to tend to infinity uniformly for t,x as $y \rightarrow \pm\infty$. Moreover, we assume that

$$\dot{V}_i(t,x,y) = \overline{\lim_{h \rightarrow 0^+}} \frac{1}{h}\{V_i(t+h,x+hy,y+hf(t,x,y))-V_i(t,x,y)\} \leq 0 \quad (5.8)$$

and

$$\dot{W}_i(t,x,y) = \overline{\lim_{h \rightarrow 0^+}} \frac{1}{h}\{W_i(t+h,x+hy,y+hf(t,x,y))-W_i(t,x,y)\} \geq 0 \quad (5.9)$$

in the interiors of their domains of definition.

Then there exists a solution $x(t)$ of (5.1) satisfying the condition

$$x(a) = \phi(a) \quad \text{and} \quad x(b) = \phi(b).$$

Proof. We can assume that $K > L$, $K > |\alpha'(t)|$ and $K > |\beta'(t)|$. Choose an $M > 0$ so large that $K \leq M$ and

$$\max\{V_1(t,x,K); \ a \leq t \leq b, \ \alpha(t) \leq x \leq \phi(t)\}$$
$$< \min\{V_1(t,x,M); \ a \leq t \leq b, \ \alpha(t) \leq x \leq \phi(t)\},$$

$$\max\{V_2(t,x,-K); \ a \leq t \leq b, \ \phi(t) \leq x \leq \beta(t)\}$$
$$< \min\{V_2(t,x,-M); \ a \leq t \leq b, \ \phi(t) \leq x \leq \beta(t)\}$$

and similarly for $W_i(t,x,y)$. For this M, construct a function $f^*(t,x,y)$ and consider the equation (5.5). Then (5.5) has a solution $x(t)$ such that $x(a) = \phi(a)$, $x(b) = \phi(b)$ and $\alpha(t) \leq x(t) \leq \beta(t)$. Now suppose $x(t_1) < \phi(t_1)$ at some t_1. Then there are $t_2, t_3, t_2 < t_3$, such that

$$x(t_2) = \phi(t_2), \ x(t_3) = \phi(t_3), \ x(t) < \phi(t) \quad \text{for} \quad t_2 < t < t_3.$$

We shall show that $x'(t) \leq M$ on this interval. Suppose $x'(t) > M$ at some t. Then there exist $t_4, t_5, t_4 < t_5$, such that $x'(t_4) = K$, $x'(t_5) = M$ and

$$K < x'(t) < M \quad \text{for} \quad t_4 < t < t_5, \quad (5.10)$$

because $x'(t_2) < L$. Moreover (5.10) implies $\alpha(t) < x(t) < \phi(t)$ on
(t_4, t_5). Considering $V_1(t,x(t),x'(t))$, we have $V_1(t_4,x(t_4),x'(t_4)) \geq$
$V_1(t_5,x(t_5),x'(t_5))$. However, since $x'(t_4) = K$ and $x'(t_5) = M$,
$V_1(t_4,x(t_4),x'(t_4)) < V_1(t_5,x(t_5),x'(t_5))$. This contradiction shows
that $x'(t) \leq M$. By using $W_2(t,x,y)$, we can see that $x'(t) \geq -M$.
In the case where we have $x(t) > \phi(t)$ at some t, we can also see
that $|x'(t)| \leq M$ by using V_2 and W_1. Thus $x(t)$ is a solution
of (5.1) such that $x(a) = \phi(a)$ and $x(b) = \phi(b)$.

Corollary 5.1. Let $\phi(t) \equiv \alpha(t)$ and assume that there exist
Liapunov functions $V_2(t,x,y)$ and $W_1(t,x,y)$ in Theorem 5.1. Then
there exists a solution $x(t)$ of (5.1) such that

$$x(a) = \alpha(a) \quad and \quad x(b) = \alpha(b).$$

Similarly, let $\phi(t) \equiv \beta(t)$ and assume the existence of
$V_1(t,x,y)$ and $W_2(t,x,y)$. Then the equation (5.1) has a solution
$x(t)$ which satisfies $x(a) = \beta(a)$ and $x(b) = \beta(b)$.

Now we assume that $V_1(t,x,y)$ and $W_1(t,x,y)$ are defined
on $a \leq t \leq b$, $\alpha(t) \leq x \leq \beta(t)$, $y \geq K$, where $K > 0$ can be large,
and $V_2(t,x,y)$, $W_2(t,x,y)$ are defined on $a \leq t \leq b$, $\alpha(t) \leq x \leq \beta(t)$,
$y \leq -K$. Moreover, we suppose that V_i, W_i, $i = 1,2$, tend to infinity
uniformly for t,x as $y \to \pm\infty$ and that

$$\dot{V}_i(t,x,y) \leq 0, \quad \dot{W}_i(t,x,y) \geq 0, \quad i = 1,2,$$

in the interiors of their domains of definition.

Theorem 5.2. Assume that $\alpha(a) = \beta(a)$ and that there exist
Liapunov functions $V_1(t,x,y)$ and $V_2(t,x,y)$ satisfying the conditions above. Then there exists a solution $x(t)$ of (5.1) passing
through $(a,\alpha(a))$ and an arbitrary point in D.

Proof. We can assume the end point to be (b,B), $\alpha(b) \leq B \leq \beta(b)$. Moreover, we can assume that $|\alpha'(t)| < K$, $|\beta'(t)| < K$. Choose an $M > 0$ so large that $K \leq M$,

$$\max\{V_1(t,x,K); \; a \leq t \leq b, \; \alpha(t) \leq x \leq \beta(t)\}$$
$$< \min\{V_1(t,x,M); \; a \leq t \leq b, \; \alpha(t) \leq x \leq \beta(t)\}$$

and

$$\max\{V_2(t,x,-K); \; a \leq t \leq b, \; \alpha(t) \leq x \leq \beta(t)\}$$
$$< \min\{V_2(t,x,-M); \; a \leq t \leq b, \; \alpha(t) \leq x \leq \beta(t)\}.$$

By the same argument as in the proof of Theorem 5.1, we can see that $x'(t) < M$, by using $V_1(t,x,y)$, and that $x'(t) > -M$, by using $V_2(t,x,y)$. Thus we can show the existence of a solution.

Remark. Assuming $\alpha(b) = \beta(b)$ and using $W_i(t,x,y)$, $i = 1,2$, we have a solution of (5.1) for right end fixed.

Corollary 5.2. Assume that $\alpha(a) = \beta(a)$. If $f(t,x,y) \leq 0$ for $y > K$ and $f(t,x,y) \geq 0$ for $y < -K$, then there exists a solution of (5.1) passing through $(a,\alpha(a))$ and an arbitrary point in D.

If we consider functions $V_1 = y^2$, $V_2 = y^2$, the conclusion follows immediately from Theorem 5.2.

CHAPTER II

STABILITY AND BOUNDEDNESS

6. Stability of a Solution

Consider a system of differential equations

$$x' = f(t,x) \qquad (' = \frac{d}{dt}).\qquad\qquad (6.1)$$

Suppose that $f(t,x) \in C(I \times D, R^n)$, where $I = [0,\infty)$ and D is a connected open set in R^n. Let F be a class of solutions of (6.1) which remain in D and let $x_0(t)$ be an element of F. Setting $x = y + x_0(t)$, the system (6.1) is transformed into

$$y' = f(t,y+x_0(t)) - f(t,x_0(t)).\qquad\qquad (6.2)$$

If we denote by $g(t,y)$ the right-hand side of (6.2), clearly $g(t,0) \equiv 0$ and the zero solution $y(t) \equiv 0$ of (6.2) corresponds to $x_0(t)$. Therefore it is sufficient to discuss the stability of $y(t) \equiv 0$ of (6.2) in place of $x_0(t)$. For this reason, we assume that $f(t,0) \equiv 0$ and that D is a domain such that $|x| < H, H > 0$.

Definition 6.1. The zero solution $x(t) \equiv 0$ of (6.1) is stable, if for any $\varepsilon > 0$ and any $t_0 \in I$, there exists a $\delta(t_0,\varepsilon) > 0$ such that $|x_0| < \delta(t_0,\varepsilon)$ implies $|x(t,t_0,x_0)| < \varepsilon$ for all $t \geq t_0$, where $x(t,t_0,x_0)$ denotes a solution of (6.1) through the point (t_0,x_0).

Definition 6.2. The zero solution of (6.1) is uniformly stable, if the δ in Definition 6.1 is independent of t_0.

Thus, when we consider a special solution $\phi(t)$ defined on I, $\phi(t)$ is uniformly stable, if for any $\varepsilon > 0$ and any $t_0 \in I$, there exists a $\delta(\varepsilon) > 0$ such that if $|\phi(t_0)-x_0| < \delta(\varepsilon)$, then

$|\phi(t)-x(t,t_0,x_0)| < \varepsilon$ for all $t \geq t_0$.

It is clear that if the zero solution of (6.1) is stable, the solution of (6.1) through the point $(t_0,0)$, $t_0 \varepsilon I$, is unique to the right. However, it is noted that even if the zero solution is stable, solutions starting near the origin are not necessarily unique. For example, consider a scalar equation $x' = f(x)$, where

$$f(x) = \begin{cases} -2\pi(2^{-k}-x)^{\frac{1}{2}}(x-2^{-k-1})^{\frac{1}{2}} & (2^{-k-1} \leq x \leq 2^{-k}), \quad k = 0,1,\ldots \\ -x & (-1 \leq x \leq 0). \end{cases}$$

Then $f(x)$ is continuous on $-1 \leq x \leq 1$, and for each $k = 0,1,\ldots$ and each $t_0 \geq 0$, there are many solutions through $(t_0,2^{-k})$, one of which is $x(t) \equiv 2^{-k}$ and another of which is

$$x(t) = 2^{-k}-2^{-k-1}\sin^2\pi(t-t_0) \quad \text{for} \quad t_0 \leq t \leq t_0+\frac{1}{2}.$$

As is seen from the following example, stability does not necessarily imply uniform stability.

Example 6.1. Consider a scalar linear equation

$$x' = (6t \sin t-2t)x. \qquad (6.3)$$

The solution $x(t,t_0,x_0)$ through (t_0,x_0) is

$$x(t,t_0,x_0) = x_0\exp\{6 \sin t-6t \cos t-t^2-6 \sin t_0+6t_0\cos t_0+t_0^2\}.$$

For $T > 6$, if $t \geq t_0+T, t_0 \geq 0$, we have

$$|x(t,t_0,x_0)| \leq |x_0|\exp\{12 + (t+t_0)(6-t+t_0)\}$$
$$\leq |x_0|\exp\{12 + T(6-T)\}. \qquad (6.4)$$

Therefore we can see that $|x(t,t_0,x_0)| \leq |x_0|M$ for some constant M which may depend on t_0. Thus the zero solution of (6.3) is stable. Now consider the solution $x(t,2k\pi,x_0)$ through $(2k\pi,x_0)$. Then

$$x((2k+1)\pi,2k\pi,x_0) = x_0 \exp\{(4k+1)\pi(6-\pi)\},$$

and hence $x((2k+1)\pi,2k\pi,x_0) \to \infty$ as $k \to \infty$ if $x_0 > 0$. Therefore the zero solution is not uniformly stable.

Example 6.2. Consider a scalar autonomous equation

$$x' = f(x), \tag{6.5}$$

where

$$f(x) = \begin{cases} -x & (x \geq 0) \\ x & (x < 0). \end{cases}$$

Then the solution $x = e^{-t}$ is stable, but it is not uniformly stable.

We now observe a periodic system

$$x' = f(t,x), \quad f(t+\omega,x) = f(t,x), \quad \omega > 0, \tag{6.6}$$

where $f(t,x) \in C(I \times D, R^n)$ and $f(t,0) \equiv 0$. This system has the following property.

Theorem 6.1. If the zero solution of (6.6) is stable, then it is uniformly stable.

For the proof, see [80].

By Theorem 6.1 and Definitions, for a periodic system the stability and the uniform stability of the zero solution are equivalent. However, as Example 6.2 shows, the stability and the uniform stability of a nontrivial solution of a periodic system are not necessarily equivalent.

In an almost periodic system, the stability of the zero solution is not necessarily equivalent to the uniform stability. Conley and Miller [13] constructed an almost periodic function $f(t)$ which has the properties that

(i) $\int_0^T f(t)\,dt \to \infty$ as $T \to \infty$,

(ii) there exist real sequences $\{t_n\}$, $\{T_n\}$ such that $t_n \to \infty$, $T_n \to \infty$ as $n \to \infty$ and that

$$\int_{t_n}^{t_n+T_n} f(t)\,dt < -n \quad \text{for} \quad n = 1,2,\dots \ .$$

Consider a scalar linear equation

$$x' = -f(t)x. \tag{6.7}$$

The solutions of (6.7) are $x(t,t_0,x_0) = x_0 \exp\!\left(-\int_{t_0}^t f(s)\,ds\right)$. By the property (i),

$$\exp\!\left(-\int_{t_0}^t f(s)\,ds\right) \to 0 \quad \text{as} \quad t \to \infty,$$

and hence there exists a constant M such that $\exp\!\left(-\int_{t_0}^t f(s)\,ds\right) \le M$, where M may depend on t_0. Thus we can see that the zero solution of (6.7) is stable. However, if we consider the solution of (6.7) through (t_n,x_0), $x_0 > 0$, then

$$\begin{aligned} x(t_n+T_n,t_n,x_0) &= x_0 \exp\!\left(-\int_{t_n}^{t_n+T_n} f(s)\,ds\right) \\ &\ge x_0 e^n. \end{aligned}$$

This shows that the zero solution of (6.7) is not uniformly stable.

Theorem 6.2. Suppose that there exists a Liapunov function $V(t,x)$ defined on $I \times D$ which satisfies the following conditions;

(i) $V(t,0) \equiv 0$,

(ii) $a(|x|) \le V(t,x)$, where $a(r)$ is a continuous positive definite function,

(iii) $\dot{V}_{(6.1)}(t,x) \le 0$.

Then the zero solution of (6.1) is stable.

Proof. Corresponding to any $\varepsilon > 0$, $\varepsilon < H$, we have $a(\varepsilon) \leq$ $V(t,x)$ for $t \in I$ and x such that $|x| = \varepsilon$. For a fixed $t_0 \in I$, we can choose a $\delta(t_0,\varepsilon) > 0$ such that $|x_0| < \delta(t_0,\varepsilon)$ implies $V(t_0,x_0) < a(\varepsilon)$, because $V(t_0,0) \equiv 0$ and $V(t,x)$ is continuous. Suppose that a solution $x(t,t_0,x_0)$ of (6.1) such that $|x_0| < \delta(t_0,\varepsilon)$ satisfies $|x(t_1,t_0,x_0)| = \varepsilon$ at some $t_1 \geq t_0$. Then, by (iii), $V(t_1,x(t_1,t_0,x_0)) \leq V(t_0,x_0)$, and hence

$$a(\varepsilon) \leq V(t_1,x(t_1,t_0,x_0)) \leq V(t_0,x_0) < a(\varepsilon).$$

This is a contradiction, and therefore, if $|x_0| < \delta(t_0,\varepsilon)$, then $|x(t,t_0,x_0)| < \varepsilon$ for all $t \geq t_0$. This shows that the zero solution is stable.

Theorem 6.3. If condition (ii) in Theorem 6.2 is replaced by

   (ii)' $a(|x|) \leq V(t,x) \leq b(|x|)$, where $a(r)$ and $b(r)$ are
          continuous, positive definite,

then the zero solution of (6.1) is uniformly stable.

By choosing a $\delta(\varepsilon) > 0$ so that $b(\delta) < a(\varepsilon)$, we can prove that if $|x_0| < \delta(\varepsilon)$ and $t_0 \in I$, then $|x(t,t_0,x_0)| < \varepsilon$ for all $t \geq t_0$.

Example 6.3. Consider Liénard's equation

$$x'' + f(x)x' + g(x) = 0, \tag{6.8}$$

where $f(x)$ and $g(x)$ are continuous on $x \in R^1$. Suppose that

   (i) $g(x)F(x) > 0$ for $x \neq 0$, where $F(x) = \int_0^x f(u)\,du$,

   (ii) $xg(x) > 0$ for $x \neq 0$,

   (iii) $G(x) = \int_0^x g(u)\,du \to \infty$ as $|x| \to \infty$.

Consider an equivalent system

$$x' = y-F(x), \quad y' = -g(x) \tag{6.9}$$

and a Liapunov function $V(t,x,y) = G(x) + y^2/2$.

Clearly V satisfies condition (ii)' in Theorem 6.3. Moreover, we have

$$\dot{V}_{(6.9)}(t,x,y) = g(x)(y-F(x)) + y(-g(x))$$

$$= -g(x)F(x)$$

$$\leq 0,$$

and hence, by Theorem 6.3, the solution $x(t) \equiv 0$, $y(t) \equiv 0$ of (6.9) is uniformly stable.

7. <u>Asymptotic Stability of a Solution</u>

Consider the system

$$x' = f(t,x), \tag{7.1}$$

where $f(t,x) \in C(I \times D, R^n)$, $D = \{x; |x| < H\}$, and $f(t,0) \equiv 0$.

<u>Definition 7.1.</u> The zero solution of (7.1) is <u>asymptotically stable</u>, if it is stable and if there exists a $\delta_0(t_0) > 0$ such that $|x_0| < \delta_0(t_0)$ implies that $x(t,t_0,x_0) \to 0$ as $t \to \infty$.

<u>Definition 7.2.</u> The zero solution of (7.1) is <u>quasi-equi-asymptotically stable</u>, if for any $\varepsilon > 0$ and any $t_0 \in I$, there exists a $\delta_0(t_0) > 0$ and a $T(t_0,\varepsilon) > 0$ such that if $|x_0| < \delta_0(t_0)$, then $|x(t,t_0,x_0)| < \varepsilon$ for all $t \geq t_0+T(t_0,\varepsilon)$.

<u>Definition 7.3.</u> The zero solution of (7.1) is <u>equi-asymptotically stable</u>, if it is stable and is quasi-equiasymptotically stable.

Here it is noticed that if the zero solution is the unique solution of (7.1) through $(0,0)$, then the quasi-equiasymptotic stability of the zero solution implies the stability, and consequently

the zero solution is equiasymptotically stable. Moreover, it is clear

that for a scalar equation, asymptotic stability is equivalent to

equiasymptotic stability.

For a linear system

$$x' = A(t)x, \qquad\qquad (7.2)$$

where   A(t)   is an   n × n   continuous matrix function on   I, we have

the following theorem.

Theorem 7.1.   If the solution   x(t) ≡ 0   of (7.2) is asymptot-

ically stable, then it is equiasymptotically stable.

Proof.   Let   X(t)   be a fundamental matrix of (7.2).   Then the

solution of (7.2) through (t_0, x_0) has the form $x(t, t_0, x_0) =$

$X(t) X^{-1}(t_0) x_0$.   Since the asymptotic stability implies that all ele-

ments of   X(t)   tend to zero   as   t → ∞, there exists a $T(t_0, \varepsilon) > 0$

such that $|X(t) X^{-1}(t_0)| < \varepsilon$, where   ε   is given.   Therefore, if

$|x_0| < 1$ and $t \geq t_0 + T(t_0, \varepsilon)$, we have

$$|x(t, t_0, x_0)| \leq |X(t) X^{-1}(t_0)| < \varepsilon,$$

which shows that   x(t) ≡ 0   is equiasymptotically stable.

Theorem 7.2.   If the system (7.1) is periodic in   t   of period

ω > 0, that is,   f(t+ω,x) = f(t,x)   for all   t ε I   and all   x ε D,

asymptotic stability of the zero solution implies equiasymptotic

stability.

For the proof, see [80].

The following example shows that the asymptotic stability is

not equivalent to the equiasymptotic stability.

Example 7.1 [45].   Consider the following system of order two which is given in polar coordinates

$$r' = r \frac{g'(t,\theta)}{g(t,\theta)} \quad , \quad \theta' = 0, \tag{7.3}$$

where $g'(t,\theta)$ denotes the derivative of $g(t,\theta)$ with respect to t and

$$g(t,\theta) = \frac{\sin^4\theta}{\sin^4\theta + (1-t\sin^2\theta)^2} + \frac{1}{1+\sin^4\theta} \cdot \frac{1}{1+t^2} \cdot \tag{7.4}$$

The solution of (7.3) which satisfies $r = r_0$, $\theta = \theta_0$ at $t = t_0$ is then

$$r = r_0 \frac{g(t,\theta_0)}{g(t_0,\theta_0)} \quad , \quad \theta = \theta_0.$$

If $\theta_0 = k\pi$, the solution is

$$r = r_0 \frac{1+t_0^2}{1+t^2} \quad , \quad \theta = k\pi.$$

If $\theta_0 \neq k\pi$, setting $\tau = \frac{1}{\sin^2\theta_0}$,

$$\begin{cases} r = r_0 \frac{1}{g(t_0,\theta_0)} \left\{ \frac{1}{1+(\tau-t)^2} + \frac{\tau^2}{1+\tau^2} \cdot \frac{1}{1+t^2} \right\} \\ \theta = \theta_0 \cdot \end{cases} \tag{7.5}$$

It is clear that $r \to 0$ as $t \to \infty$, and hence the solution $r = 0$ is asymptotically stable. However, if θ_0 is very near $k\pi$ and $t_0 = 0$, the solution (7.5) will have values $r > r_0$ at $t = \tau$ which are as large as we please. Therefore the solution is not equi-asymptotically stable.

Now we shall give the definition of uniformly asymptotic stability.

Definition 7.4. The zero solution of (7.1) is <u>quasi-uniformly</u>

asymptotically stable, if the δ_0 and the T in Definition 7.2 are
independent of t_0.

Definition 7.5. The zero solution of (7.1) is uniformly
asymptotically stable, if it is uniformly stable and is quasi-uniformly
asymptotically stable.

Definition 7.6. The zero solution of (7.1) is exponentially
asymptotically stable, if there exists a $\lambda > 0$ and, given any
$\epsilon > 0$, there exists a $\delta(\epsilon) > 0$ such that $|x_0| < \delta(\epsilon)$ implies

$$|x(t,t_0,x_0)| \leq \epsilon e^{-\lambda(t-t_0)} \quad \text{for all} \quad t \geq t_0.$$

Example 6.1 shows that the zero solution of (6.3) is quasi
uniformly asymptotically stable, because by (6.4), we have
$|x(t,t_0,x_0)| < \epsilon$ for all $t \geq t_0+T$, if $|x_0| \leq 1$ and
$\exp\{12 + T(6-T)\} < \epsilon$. However, as was seen, the zero solution is not
uniformly stable. Therefore, the zero solution of (6.3) is equiasymptotically stable, but it is not uniformly asymptotically stable.

Moreover, the zero solution of the following example is uniformly stable, but it is not uniformly asymptotically stable.

Example 7.2. Consider a scalar linear equation

$$x' = -\frac{x}{1+t} . \tag{7.6}$$

The solution through (t_0,x_0) is $x(t,t_0,x_0) = \frac{1+t_0}{1+t}x_0$. For
$|x_0| \leq 1$, if we want to have $|x(t,t_0,x_0)| < \epsilon$ for $t \geq t_0+T$, T
must be greater than $(t_0+1)(\frac{1}{\epsilon} - 1)$.

Theorem 7.3. If the zero solution of the linear system (7.2)
is uniformly asymptotically stable, then it is exponentially asymptotically stable.

For the proof, see [80]. In this case, there exist two positive constants K and λ for which $|x(t,t_0,x_0)| \le K e^{-\lambda(t-t_0)} |x_0|$ for all $t \ge t_0$ and all $x_0 \in R^n$.

Example 7.3. Consider a simple equation $x' = -x^3$. Then the solution through (t_0,x_0) is

$$x(t,t_0,x_0) = x_0 [1+2x_0^2 (t-t_0)]^{-1/2}.$$

It is easy to see that the zero solution is uniformly asymptotically stable, but it is not exponentially asymptotically stable.

Now consider a periodic system

$$x' = f(t,x), \tag{7.7}$$

where $f(t,x) \in C(I \times D, R^n)$, $f(t,0) \equiv 0$ for all $t \in I$ and $f(t+\omega,x) = f(t,x)$, $\omega > 0$, for all t and $x \in D$. Then we have the following properties.

Theorem 7.4. If the zero solution of (7.7) is asymptotically stable, then it is uniformly asymptotically stable.

For the proof, see [80].

Lemma 7.1. More generally, consider an almost periodic system

$$x' = f(t,x), \tag{7.8}$$

where $f(t,x) \in C(R \times D, R^n)$ and $f(t,x)$ is almost periodic in t uniformly for $x \in D$ (note that we do not assume $f(t,0) \equiv 0$).

Suppose that for any $t_0 \in I$ and any $\varepsilon > 0$, there exists a $\delta_0(t_0) > 0$ and a $T(t_0,\varepsilon) \ge 0$ such that $|x_0| < \delta_0(t_0)$ implies $|x(t,t_0,x_0)| < \varepsilon$ for all $t \ge t_0 + T(t_0,\varepsilon)$. Then the system has the zero solution defined on $[0,\infty)$.

Remark. For the existence of the zero solution, it is suffici-ent to assume $t_0 = 0$.

Proof. Let $z(t)$ be a solution of (7.8) such that
$|z(0)| < \delta_0(0)$. Then $z(t) \to 0$ as $t \to \infty$. For each $t \geq 0$,

$$z(t+k) \to 0 \quad \text{as} \quad k \to \infty, \tag{7.9}$$

where k is a positive integer. Clearly $y_k(t) = z(t+k)$ is a solution of

$$x' = f(t+k,x) \tag{7.10}$$

and $y_k(0) = z(k)$. Since $z(t) \to 0$ as $t \to \infty$, there is an H^*,
$H^* < H$, such that $|z(t)| \leq H^*$ for all $t \geq 0$. $f(t,x)$ is almost
periodic in t uniformly for $x \varepsilon D$ and the set $S = \{x; |x| \leq H^*\}$
is compact, and hence there is an $L > 0$ such that $|f(t,x)| \leq L$
for all $t \varepsilon I$ and $x \varepsilon S$. Therefore $\{y_k(t)\}$, $t \geq 0$, is uniformly
bounded and equicontinuous. Thus, for the sequence $\{k\}$, there exists
a subsequence $\{k_j\}$ such that

$$f(t+k_j,x) \to g(t,x) \quad \text{uniformly on} \quad R \times S,$$
$$y_{k_j}(0) = z(k_j) \to 0$$

as $j \to \infty$ and that for each fixed $\tau > 0$

$$y_{k_j}(t) \to \eta(t) \quad \text{on} \quad 0 \leq t \leq \tau,$$

where $g \varepsilon H(f)$ and $\eta(t)$ is a solution through $(0,0)$ of

$$x' = g(t,x). \tag{7.11}$$

This means that $z(t+k_j) \to \eta(t)$, $t \varepsilon [0,\tau]$, as $j \to \infty$. By (7.9),
$\eta(t) \equiv 0$ for $0 \leq t \leq \tau$. Since τ is arbitrary, the zero function
is a solution of (7.11) on I. Since $g \varepsilon H(f)$ implies $f \varepsilon H(g)$,
the zero function is also a solution of (7.8) on I.

Theorem 7.5. [66]. Consider the periodic system (7.7) without

assuming $f(t,0) \equiv 0$. Under the same assumption as in Lemma 7.1,
the system has the zero solution which is stable, and consequently,
the zero solution is uniformly asymptotically stable.

   Proof. The existence of the zero solution on I follows immediately from Lemma 7.1. We now show that it is unique to the right.
Then the zero solution will be equiasymptotically stable, and hence,
by Theorem 7.4, it is uniformly asymptotically stable.

   Let $y(t,0,0)$ denote any solution of (7.7) through $(0,0)$.
Define

$$y_k(t,0,0) = \begin{cases} 0 & \text{if } 0 \le t \le k\omega \\ y(t-k\omega,0,0) & \text{if } k\omega < t \end{cases}$$

for $k = 1,2,\ldots$. Since the zero function is a solution of (7.7),
so is $y_k(t,0,0)$ for each k. For a given $\varepsilon > 0$, consider $T(0,\varepsilon)$.
Then $|y_k(t,0,0)| < \varepsilon$ for all $t \ge T(0,\varepsilon)$ and all k. Choose an
integer $N(\varepsilon)$ so large that $N(\varepsilon)\omega \ge T(0,\varepsilon)$. Let $t \ge N(\varepsilon)\omega$. Then
we have

$$|y_{N(\varepsilon)}(t,0,0)| = |y(t-N(\varepsilon)\omega,0,0)| < \varepsilon \quad \text{for all } t \ge N(\varepsilon)\omega.$$

This means $|y(t,0,0)| < \varepsilon$ for all $t \ge 0$. Since ε was arbitrary,
$y(t,0,0) = 0$ for all $t \ge 0$. Thus the zero solution is unique to the
right. This completes the proof.

   Sell [63] has introduced other types of stability. Consider
a system

$$x' = f(t,x), \tag{7.12}$$

where $f(t,x) \in C(I \times D, R^n)$.

   Definition 7.7. A given solution $\phi(t)$ of (7.12) such that

$|\phi(t)| \leq H^*$, $H^* < H$, for all $t \geq 0$ is said to be <u>weakly uniformly</u> <u>asymptotically stable</u>, if it is uniformly stable and there is a $\delta_0 > 0$ such that if $t_0 \varepsilon I$ and $|\phi(t_0) - x_0| \leq \delta_0$, then $|x(t,t_0,x_0) - \phi(t)| \to 0$ as $t \to \infty$.

Actually, this stability is weaker than the uniformly asymptotic stability as Example 7.2 shows. The zero solution of (7.6) is uniformly stable and is weakly uniformly asymptotically stable. In fact, the zero solution of (7.6) is equiasymptotically stable. Now we shall show that weakly uniformly asymptotic stability implies equiasymptotic stability.

Theorem 7.6. If a solution $\phi(t)$ of (7.12) such that $|\phi(t)| \leq H^*$, $H^* < H$, for all $t \geq 0$ is weakly uniformly asymptotically stable, then $\phi(t)$ is equiasymptotically stable.

More precisely, it is stable and for any $\varepsilon > 0$, there exists a $T(t_0, \varepsilon) > 0$ such that if $t_0 \varepsilon I$ and $|\phi(t_0) - x_0| \leq \delta_0$, where $\delta_0 < H - H^*$, then

$$|x(t,t_0,x_0) - \phi(t)| < \varepsilon \quad \text{for all} \quad t \geq t_0 + T(t_0, \varepsilon).$$

Proof. Since $\phi(t)$ is uniformly stable, $\phi(t)$ is stable. Suppose that there is no T. Then there exists some $\varepsilon > 0$, $t_0 \geq 0$ and sequences $\{x_k\}$, $\{t_k\}$ such that $|x_k - \phi(t_0)| \leq \delta_0$, $t_k \to \infty$ as $k \to \infty$ and

$$|x(t_k,t_0,x_k) - \phi(t_k)| \geq \varepsilon. \tag{7.13}$$

On any compact interval $[t_0, t_0+N]$, the sequence $\{x(t,t_0,x_k)\}$ is uniformly bounded and equicontinuous, and hence we can find a solution $x(t,t_0,\bar{x})$ of (7.12) defined for all $t \geq t_0$, where $|\phi(t_0) - \bar{x}| \leq \delta_0$, and moreover, a subsequence of $\{x(t,t_0,x_k)\}$ tends to $x(t,t_0,\bar{x})$ uniformly on any compact interval. Since every solution tends to $\phi(t)$ as $t \to \infty$, we have at some t_1

$$|x(t_1,t_0,\overline{x}) - \phi(t_1)| < \frac{1}{2}\, \delta(\varepsilon), \qquad (7.14)$$

where $\delta(\varepsilon)$ is the one for uniform stability. Denoting by
$\{x(t,t_0,x_k)\}$ the subsequence again, if k is sufficiently large, we
have

$$|x(t_1,t_0,x_k)-x(t_1,t_0,\overline{x})| < \frac{1}{2}\, \delta(\varepsilon).$$

From this and (7.14) it follows that $|x(t_1,t_0,x_k)-\phi(t_1)| < \delta(\varepsilon)$.
Therefore, by the uniform stability of $\phi(t)$, we have

$$|x(t,t_0,x_k)-\phi(t)| < \varepsilon \quad \text{for all}\quad t \geq t_1,$$

which contradicts (7.13). This proves the theorem.

For the periodic system, we have the following theorem [82].

Theorem 7.7. Suppose that the system (7.12) is periodic in t
of period $\omega > 0$ and suppose that the system has a solution $\phi(t)$
such that $|\phi(t)| \leq H^*$, $H^* < H$, for all $t \geq 0$. If $\phi(t)$ is weakly
uniformly asymptotically stable, then it is uniformly asymptotically
stable.

Proof. Since $\phi(t)$ is uniformly stable, there exists a
$\delta_0^* > 0$ such that $t_0 \in I$ and $|\phi(t_0)-x_0| \leq \delta_0^*$ implies
$|x(t,t_0,x_0)-\phi(t)| < \dfrac{\delta_0}{2}$ for all $t \geq t_0$, where δ_0 is the one in
Definition 7.7. Suppose that for this δ_0^*, $\phi(t)$ is not uniformly
asymptotically stable. Then, for some $\varepsilon > 0$, there exist sequences
$\{k_j\}$, $\{x_{k_j}\}$ and $\{\tau_{k_j}\}$ such that $k_j \to \infty$, $\tau_{k_j} \to \infty$ as $j \to \infty$, where
k_j is a positive integer, and that

$$|x_{k_j}-\phi(k_j\omega)| < \frac{\delta_0}{2} \qquad (7.15)$$

and

$$|x(k_j\omega + \tau_{k_j}, k_j\omega, x_{k_j}) - \phi(k_j\omega + \tau_{k_j})| \geq \varepsilon. \tag{7.16}$$

Clearly, $|\phi(k_j\omega)| \leq H^*$, and hence there is a subsequence $\{m_j\}$ of $\{k_j\}$ and ϕ_0 such that $m_j \to \infty$ monotonically as $j \to \infty$ and $\phi(m_j\omega) \to \phi_0$ as $j \to \infty$. Then there exists an integer $p > 0$ such that if $j \geq p$, we have $|\phi(m_j\omega) - \phi_0| < \frac{\delta_0}{4}$. Thus, for any $j \geq p$ we have

$$|\phi(m_j\omega) - \phi(m_p\omega)| < \frac{\delta_0}{2}. \tag{7.17}$$

From (7.15) with $k_j = m_j$ and (7.17), it follows that

$$|x_{m_j} - \phi(m_p\omega)| < \delta_0. \tag{7.18}$$

By Theorem 7.6, there exists a $T(m_p\omega, \frac{\varepsilon}{2}) > 0$ such that

$$|x(t, m_p\omega, \phi(m_j\omega)) - \phi(t)| < \frac{\varepsilon}{2} \quad \text{for all} \quad t \geq m_p\omega + T(m_p\omega, \frac{\varepsilon}{2})$$

and

$$|x(t, m_p\omega, x_{m_j}) - \phi(t)| < \frac{\varepsilon}{2} \quad \text{for all} \quad t \geq m_p\omega + T(m_p\omega, \frac{\varepsilon}{2}).$$

This implies that

$$|x(t, m_p\omega, x_{m_j}) - x(t, m_p\omega, \phi(m_j\omega))| < \varepsilon \tag{7.19}$$

for all $t \geq m_p\omega + T(m_p\omega, \frac{\varepsilon}{2})$. Since ω is the period and m_j, m_p are integers, it follows from (7.19) that for any $j \geq p$

$$|x(t, m_j\omega, x_{m_j}) - \phi(t)| < \varepsilon \quad \text{for all} \quad t \geq m_j\omega + T(m_p\omega, \frac{\varepsilon}{2}).$$

This contradicts (7.16), because $T(m_p\omega, \frac{\varepsilon}{2})$ depends only on ε. This completes the proof.

For the almost periodic system (7.8), we have the following theorem [62].

Theorem 7.8. If $x(t) \equiv 0$ is a solution of (7.8) which is weakly uniformly asymptotically stable, then it is uniformly asymptotically stable.

   Proof. Since the zero solution is uniformly stable, there

exists a $\delta(\delta_0) > 0$ such that $|x_0| \leq \delta(\delta_0)$ implies

$|x(t,t_0,x_0)| < \delta_0$ for all $t \geq t_0$, $t_0 \geq 0$, where δ_0 is the number

given in Definition 7.7. Let $\varepsilon > 0$ be given. We shall now show

that there exists a $T(\varepsilon) > 0$ such that for any x_0, $|x_0| \leq \delta(\delta_0)$,

and for any $t_0 \varepsilon I$, there exists a $t_1, t_0 \leq t_1 \leq t_0 + T(\varepsilon)$, such that

$|x(t_1,t_0,x_0)| < \delta(\varepsilon)$, where $\delta(\varepsilon)$ is the one for the uniform stability

of $x(t) \equiv 0$. Then, clearly it will follow that $|x(t,t_0,x_0)| < \varepsilon$

for $t \geq t_0 + T(\varepsilon)$, which shows that the zero solution is uniformly

asymptotically stable.

   Suppose that there is no $T(\varepsilon)$. Then for each integer $k \geq 1$,

there exists an x_k and a $t_k \varepsilon I$ such that $|x_k| \leq \delta(\delta_0)$ and

$|x(t,t_k,x_k)| \geq \delta(\varepsilon)$ for all $t_k \leq t \leq t_k + k$. Letting $y_k(t) =$

$x(t+t_k,t_k,x_k)$, $y_k(t)$ is a solution of

$$x' = f(t+t_k,x) \qquad\qquad (7.20)$$

through the point $(0,x_k)$ and $\|y_k(t)\| \leq \delta(\varepsilon)$ on $0 \leq t \leq k$. Since

$|x_k| \leq \delta(\delta_0)$, $|y_k(t)| \leq \delta_0$ and $f(t,x)$ is almost periodic in t

uniformly for $x \varepsilon D$, there exist an x_0, functions $g(t,x)$, $z(t)$ and

a subsequence $\{k_j\}$ of $\{k\}$ such that

   $x_{k_j} \rightarrow x_0$,

   $f(t+t_{k_j},x) \rightarrow g(t,x)$ uniformly on $I \times \{x:|x| \leq \delta_0\}$

and

   $y_{k_j}(t) \rightarrow z(t)$ uniformly on any compact interval on I.

Clearly $z(t)$ is a solution of

$$x' = g(t,x), \qquad\qquad (7.21)$$

which is defined on I and passes through $(0,x_0)$. For fixed $t \geq 0$, there is a j sufficiently large so that

$$|y_{k_j}(t)| - |y_{k_j}(t)-z(t)| \leq |z(t)|.$$

Since $|y_{k_j}(t)| \geq \delta(\varepsilon)$ and $|y_{k_j}(t)-z(t)| < \frac{\delta(\varepsilon)}{2}$ for large j, we have

$$|z(t)| \geq \frac{\delta(\varepsilon)}{2} \quad \text{for all} \quad t \geq 0. \tag{7.22}$$

Moreover, clearly

$$|z(t)| \leq \delta_0 \quad \text{for all} \quad t \geq 0. \tag{7.23}$$

Since g is in $H(f)$, f is in $H(g)$, and hence there exists a sequence $\{\tau_k\}$ such that $\tau_k \to \infty$ as $k \to \infty$ and $g(t+\tau_k,x) \to f(t,x)$ uniformly for $t \in R$ and $x \in \{x; |x| \leq \delta_0\}$ as $k \to \infty$. If we set $\eta_k(t) = z(t+\tau_k)$, $\eta_k(t)$ is a solution through $(0,z(\tau_k))$ of

$$x' = g(t+\tau_k,x). \tag{7.24}$$

Since $|z(t)| \leq \delta_0$ for all $t \geq 0$, $\{\eta_k(t)\}$ is uniformly bounded and equicontinuous. Hence there exists a subsequence $\{\tau_{k_j}\}$ of $\{\tau_k\}$ such that

$$g(t+\tau_{k_j},x) \to f(t,x) \quad \text{uniformly for} \quad t \in R \quad \text{and} \quad |x| \leq \delta_0$$

and

$$\eta_{k_j}(t) \to \xi(t) \quad \text{on any compact interval on} \quad I,$$

where $\xi(t)$ is a solution of (7.8). For fixed $t \geq 0$, there exists a j so large that

$$|\xi(t)| \geq |\eta_{k_j}(t)| - |\eta_{k_j}(t)-\xi(t)| \geq \frac{\delta(\varepsilon)}{2} - \frac{\delta(\varepsilon)}{4} = \frac{\delta(\varepsilon)}{4}, \tag{7.25}$$

because $\tau_{k_j} > 0$ for j sufficiently large and

$|\eta_{k_j}(t)| = |z(t+\tau_{k_j})| \geq \dfrac{\delta(\varepsilon)}{2}$ by (7.22). Moreover, by (7.23),

we have $|\xi(0)| \leq \delta_0$. But this implies that $\xi(t) \to 0$ as $t \to \infty$,
which contradicts (7.25). This proves the theorem.

Theorem 7.9. Suppose that there exists a Liapunov function
$V(t,x)$ defined on $I \times D$ which satisfies the following conditions;

   (i) $a(|x|) \leq V(t,x) \leq b(|x|)$, where $a(r)$ and $b(r)$ are con-
         tinuous, positive definite,
   (ii) $\overset{\cdot}{V}_{(7.1)}(t,x) \leq -c(|x|)$, where $c(r)$ is continuous and
         positive definite.

Then the zero solution of (7.1) is uniformly asymptotically stable.

   For the proof, see [80].

Corollary. Under the same assumption as in Theorem 7.9, if
$\overset{\cdot}{V}_{(7.1)}(t,x) \leq -cV(t,x)$, where $c > 0$ is a constant, then the zero
solution $x(t) \equiv 0$ of (7.1) is uniformly asymptotically stable.

   This is an immediate consequence of Theorem 7.9. However,
applying Theorem 1.1,

$$V(t,x(t,t_0,x_0)) \leq V(t_0,x_0)e^{-c(t-t_0)}, \quad t \geq t_0,$$

and hence uniformly asymptotic stability can be easily proved. As
we shall see later, the existence of $V(t,x)$ satisfying the condition
in the Corollary is a necessary condition for uniformly asymptotic
stability and it is very useful in discussing the behavior of solutions
of perturbed systems.

   The following theorem is a sufficient condition for asymptotic
stability, which does not necessarily imply uniformly asymptotic
stability.

Theorem 7.10. Suppose that there exists a Liapunov function

$V(t,x)$ defined on $I \times D$ which satisfies the following conditions;

> (i) $V(t,0) \equiv 0$,
>
> (ii) $a(|x|) \leq V(t,x)$, where $a(r)$ is continuous, positive definite,
>
> (iii) $\dot{V}_{(7.1)}(t,x) \leq -c(|x|)$, where $c(r)$ is continuous and positive definite.

If $f(t,x)$ is bounded when x is contained in a compact set, then the zero solution of (7.1) is asymptotically stable.

Proof. By Theorem 6.2, $x(t) \equiv 0$ is stable. Therefore, for every $t_0 \varepsilon I$, there is a $\delta_0(t_0) > 0$ such that $|x_0| < \delta_0(t_0)$ implies $|x(t,t_0,x_0)| \leq H^* < H$. We shall show that every solution $x(t,t_0,x_0)$ such that $|x_0| < \delta_0(t_0)$ tends to zero as $t \to \infty$. Suppose that some solution $x(t,t_0,x_0)$ does not tend to zero as $t \to \infty$. Then for some $\varepsilon > 0$, there exists a divergent sequence $\{t_k\}$ for which $|x(t_k,t_0,x_0)| \geq \varepsilon$, where $t_0 \varepsilon I$ and $|x_0| < \delta_0(t_0)$. Since $f(t,x)$ is bounded for x such that $|x| \leq H^*$, there is a $K > 0$ such that

$$\left| \frac{d}{dt} |x(t,t_0,x_0)| \right| < K.$$

Therefore, on the intervals

$$t_k - \frac{\varepsilon}{2K} \leq t \leq t_k + \frac{\varepsilon}{2K} ,$$

we have $|x(t,t_0,x_0)| \geq \frac{\varepsilon}{2}$. We can assume that these intervals are disjoint and $t_1 - \frac{\varepsilon}{2K} > t_0$ by taking, if necessary, a subsequence of $\{t_k\}$. Since $\dot{V}_{(7.1)}(t,x) \leq -c(|x|)$, there exists a constant $\gamma > 0$ such that $V'(t,x(t,t_0,x_0)) \leq -\gamma$ on the intervals (7.26), and $V'(t,x(t,t_0,x_0)) \leq 0$ elsewhere, because $\frac{\varepsilon}{2} \leq |x(t,t_0,x_0)| \leq H^*$ on the intervals (7.26). Therefore

$$V(t_k + \frac{\varepsilon}{2K}, x(t_k + \frac{\varepsilon}{2K},t_0,x_0)) - V(t_0,x_0) < -\gamma \frac{\varepsilon}{K} k \to -\infty$$

as $k \to \infty$, which contradicts $V(t,x) \geq 0$. Thus we see that the zero solution of (7.1) is asymptotically stable.

Example 7.4. Consider the scalar equation (7.6). As was seen, the zero solution of (7.6) is asymptotically stable, but not uniformly asymptotically stable. For the equation (7.6), consider a Liapunov function $V(t,x) = (1+t)x^2$. Clearly

$$V(t,0) \equiv 0, \quad x^2 \leq V(t,x)$$

and

$$\dot{V}_{(7.6)}(t,x) = x^2 + (1+t)2x(-\frac{x}{1+t}) = -x^2.$$

Remark. In Theorem 7.10, we can not drop the condition that $f(t,x)$ is bounded.

Example 7.5. [45] Consider a linear equation

$$x' = \frac{g'(t)}{g(t)} x, \tag{7.27}$$

where

$$g(t) = \sum_{n=1}^{\infty} \frac{1}{1+n^4(t-n)^2}.$$

The series converges uniformly on $0 \leq t < \infty$ and $g(t)$ has the following properties;

$$1 < g(n),$$
$$g(t) < 2 + \sum_{n=1}^{\infty} n^{-4} = C \quad \text{for any} \quad t.$$

Moreover, it is not difficult to see that $g'(t)$ exists and is continuous.

The solution $x(t,t_0,x_0)$ of (7.27) is given by

$$x(t,t_0,x_0) = \frac{g(t)}{g(t_0)} x_0.$$

Since $g(n) > 1$, it is clear that $x(t) \equiv 0$ is not asymptotically stable. For any $t > 0$,

$$\int_0^t \frac{dt}{1+n^4(t-n)^2} \leq \frac{1}{n^2} \pi,$$

and hence the series can be integrated termwise on $0 \leq t < \infty$ and $\int_0^\infty g(t)dt < \infty$. Since $g(t)$ is bounded, $g^2(t)$ is integrable on I. Consider a Liapunov function

$$V(t,x) = \frac{x^2}{g^2(t)} \{c^2 + \int_t^\infty g^2(t)dt\}.$$

Then we have $V(t,0) \equiv 0$ and

$$x^2 \leq V(t,x), \quad \dot{V}_{(7.27)}(t,x) = -x^2.$$

Thus this function $V(t,x)$ satisfies the conditions in Theorem 7.10. However, $\frac{g'(t)}{g(t)}$ is unbounded. This will be shown in the following way. Since $g(n) > 1$ and $g(t)$ is continuous, if $|g'(t)|$ is bounded, say $|g'(t)| \leq K$ for all $t \geq 0$, we have $g(t) \geq \frac{1}{2}$ on $n \leq t \leq n + \frac{1}{2K}$, and hence $g(t)$ is not integrable. Therefore $|g'(t)|$ is unbounded. Since $\frac{|g'(t)|}{g(t)} \geq \frac{|g'(t)|}{c}$, $\frac{|g'(t)|}{g(t)}$ is unbounded.

8. Boundedness of Solutions

Consider a system of differential equations

$$x' = f(t,x), \qquad\qquad (8.1)$$

where $f(t,x) \varepsilon C(I \times R^n, R^n)$.

Definition 8.1. A solution $x(t,t_0,x_0)$ of (8.1) is **bounded**, if there exists a $\beta > 0$ such that $|x(t,t_0,x_0)| < \beta$ for all $t \geq t_0$, where β may depend on each solution.

Definition 8.2. The solutions of (8.1) are **equi-bounded**, if for any $\alpha > 0$ and any $t_0 \varepsilon I$, there exists a $\beta(t_0,\alpha) > 0$ such that $|x_0| \leq \alpha$ implies $|x(t,t_0,x_0)| < \beta(t_0,\alpha)$ for all $t \geq t_0$.

Definition 8.3. The solutions of (8.1) are **uniformly bounded**, if the β in Definition 8.2 is independent of t_0.

It is evident that a linear transformation of coordinates does not affect the boundedness properties as well as the stability properties. However, a general transformation of coordinates will affect those properties.

For a linear system, the following properties can be easily demonstrated.

Theorem 8.1. If all solutions are bounded, then they are equi-bounded. Moreover, stability and boundedness (consequently equi-boundedness) are equivalent, and uniform stability and uniform boundedness are equivalent.

For the proof, refer to [42] and [73].

For the periodic system

$$x' = f(t,x), \quad f(t+\omega,x) = f(t,x), \quad \omega > 0, \qquad (8.2)$$

where $f(t,x) \in C(I \times R^n, R^n)$, we have the following property.

Theorem 8.2. If the solutions of (8.2) are equi-bounded, then they are uniformly bounded.

Proof. For a given $\alpha > 0$, consider solutions starting from (t_0, x_0) such that $0 \leq t_0 < \omega$ and $|x_0| \leq \alpha$. Since every solution is continuable to $t = \omega$, there exists a $\beta(\alpha) > 0$ by which the solutions considered are bounded on $[t_0, \omega]$. The solutions are equi-bounded, and hence there exists a $\gamma(\beta) > 0$ such that if $|x_0| \leq \beta$, then $|x(t,\omega,x_0)| < \gamma$ for all $t \geq \omega$. Thus, if $0 \leq t_0 < \omega$ and $|x_0| \leq \alpha$, we have $|x(t,t_0,x_0)| < \gamma$ for all $t \geq t_0$. From the periodicity of $f(t,x)$, it follows that $t_0 \in I$ and $|x_0| \leq \alpha$ imply $|x(t,t_0,x_0)| < \gamma$ for all $t \geq t_0$. This proves uniform boundedness.

Remark. The existence of $\beta(\alpha)$ in the proof follows from the following lemma.

Lemma 8.1. Suppose that $f(t,x)$ of (8.1) is continuous on $[0,T] \times R^n$ and let K be a compact set in $[0,T] \times R^n$. If every solution $x(t,t_0,x_0)$ of (8.1) through $(t_0,x_0) \varepsilon K$ is continuable to $t = T$, then there exists a $\beta(K) > 0$ such that $|x(t,t_0,x_0)| < \beta(K)$ for all $t \varepsilon [t_0,T]$.

For the proof, see [80].

It is clear that for a scalar equation, the boundedness of solutions implies the equi-boundedness. For a linear system, equi-boundedness does not necessarily imply uniform boundedness as is seen from Example 6.1.

As the following example shows, the boundedness of solutions does not necessarily imply the equi-boundedness of solutions.

Example 8.1. Consider the system of order two which is given in polar coordinates

$$r' = r \frac{g'(t,\theta)}{g(t,\theta)} , \quad \theta' = 0, \tag{8.3}$$

where $g'(t,\theta)$ is the derivative with respect to t and $g(t,\theta)$ is given by

$$g(t,\theta) = \frac{(1+t)\sin^4\theta}{\sin^4\theta+(1-t\sin^2\theta)^2} + \frac{1}{1+\sin^4\theta} \cdot \frac{1}{1+t^2} .$$

The solution such that $r = r_0$, $\theta = \theta_0$ at $t = 0$ is

$$r = r_0 \frac{g(t,\theta_0)}{g(t_0,\theta_0)}, \quad \theta = \theta_0.$$

If $\theta_0 = k\pi$ (k:integer), the solution is

$$r = r_0 \frac{1+t_0^2}{1+t^2} , \quad \theta = k\pi,$$

and if $\theta_0 \neq k\pi$, the solution may be written as

$$r = r_0 \left\{ \frac{1+t}{1+(t-\tau)^2} + \frac{\tau^2}{1+\tau^2} \cdot \frac{1}{1+t^2} \right\} \cdot \frac{1}{g(t_0, \theta_0)} , \quad \theta = \theta_0 ,$$

where $\tau = \dfrac{1}{\sin^2 \theta_0}$. It is clear that every solution is bounded, and if

θ_0 is very near $k\pi$, the solution will have a great value r for

$t = \tau$ which is as large as we please, and hence the solutions are not

equi-bounded.

Moreover, even for an autonomous system, the solutions are not

necessarily equi-bounded, though all solutions are bounded as the

following example due to Yorke shows.

Example 8.2. Consider a system

$$\begin{aligned} x' &= 0 \\ y' &= -z|x| \\ z' &= yx^2 \end{aligned} \qquad\qquad (8.4)$$

and the solution of (8.4) through $(0, x_0, y_0, z_0)$. Clearly, if $x_0 = 0$,

the solution is

$$x = 0, \quad y = y_0, \quad z = z_0 .$$

If $x_0 \neq 0$, the solution is

$$\begin{aligned} x &= x_0 \\ y &= y_0 \cos \sqrt{|x_0|^3}\, t - \frac{z_0}{\sqrt{|x_0|}} \sin \sqrt{|x_0|^3}\, t \\ z &= y_0 \sqrt{|x_0|} \sin \sqrt{|x_0|^3}\, t + z_0 \cos \sqrt{|x_0|^3}\, t. \end{aligned}$$

Thus we see that every solution is bounded. However, $|y|$ will have

a great value which is as large as we please if $|x_0|$ is small enough.

Therefore the solutions are not equi-bounded.

Definition 8.4. The solutions of (8.1) are <u>ultimately bounded</u> <u>for bound B</u>, if there exists a $B > 0$ and a $T > 0$ such that for every solution $x(t,t_0,x_0)$ of (8.1), $|x(t,t_0,x_0)| < B$ for all $t \geq t_0+T$, where B is independent of the particular solution while T may depend on each solution.

Definition 8.5. The solutions of (8.1) are <u>equi-ultimately</u> <u>bounded for bound B</u>, if there exists a $B > 0$ and if corresponding to any $\alpha > 0$ and $t_0 \varepsilon I$, there exists a $T(t_0,\alpha) > 0$ such that $|x_0| \leq \alpha$ implies that $|x(t,t_0,x_0)| < B$ for all $t \geq t_0+T(t_0,\alpha)$.

Definition 8.6. The solutions of (8.1) are <u>uniformly ulti-</u> <u>mately bounded for bound B</u>, if the T in Definition 8.5 is independent of t_0.

These concepts are actually different concepts. This is clear if we observe Examples 7.1 and 7.2. It is evident that the solutions of (8.1) are equi-bounded if they are equi-ultimately bounded. For the linear system and the periodic system, we have the following properties.

Theorem 8.3. If the solutions of a linear system are ultimately bounded, they are equi-ultimately bounded.

Theorem 8.4. If the solutions of the periodic system (8.2) are equi-ultimately bounded, then they are uniformly ultimately bounded.

For the periodic system (8.2), we have the following result [58].

Theorem 8.5. Assume that the solution of (8.2) is unique for the initial value problem. If the solutions of (8.2) are ultimately bounded for bound B, then the solutions of (8.2) are uniformly bounded. This implies that the solutions of (8.2) are uniformly

ultimately bounded.

   Proof. It is sufficient to show that the solutions of (8.2)

are equi-bounded. Suppose that the solutions are not equi-bounded.

Then there exist an $\alpha > 0$, $t_0 \geq 0$, sequences $\{x_k\}$ and $\{\tau_k\}$ such

that $|x_k| \leq \alpha$, $\tau_k \geq t_0$ and $|x(\tau_k, t_0, x_k)| \geq k$, where we can assume

that $\alpha > B$ and $k > \alpha$. Moreover, there exists a t_k such that

$$|x(t_k, t_0, x_k)| = \alpha$$

and

$$\alpha < |x(t, t_0, x_k) \quad \text{for} \quad t_k < t \leq \tau_k.$$

   Let $m_k \geq 0$ be an integer such that $t_k = m_k \omega + \sigma_k$, $0 \leq \sigma_k < \omega$,

and set $\tau_k = m_k \omega + \tau_k'$ and $y_k = x(t_k, t_0, x_k)$. Then $|y_k| = \alpha$ and

$x(t, t_0, x_k) = x(t, t_k, y_k)$ for $t \geq t_k$. Since the system is periodic

of period ω, there is a solution of (8.2) such that

$$x(t, \sigma_k, y_k) = x(t + m_k \omega, t_k, y_k) \quad \text{for} \quad t \geq \sigma_k.$$

Therefore $|x(\tau_k', \sigma_k, y_k)| \geq k$, $|x(\sigma_k, \sigma_k, y_k)| = \alpha$ and $\alpha < |x(t, \sigma_k, y_k)|$

for $\sigma_k < t \leq \tau_k'$. There are subsequences $\{y_{k_j}\}$ and $\{\sigma_{k_j}\}$ such

that $y_{k_j} \to y_0$, $\sigma_{k_j} \to 0$ as $j \to \infty$, and we have $|y_0| = \alpha$,

$0 \leq \sigma_0 \leq \omega$. Let $x(t, \sigma_0, y_0)$ be the solution of (8.2) through (σ_0, y_0).

Then there exists a $T = T(\sigma_0, y_0) > 0$ such that

$$|x(t, \sigma_0, y_0)| < B \quad \text{for all} \quad t \geq T.$$

Since we assume the uniqueness, T depends only on σ_0, y_0. By the

uniqueness, if j is sufficiently large, $x(t, \sigma_{k_j}, y_{k_j})$ remains in a

small neighborhood of $x(t, \sigma_0, y_0)$ on the interval $\sigma_{k_j} \leq t \leq T$, and

hence τ_{k_j}' cannot be less than T. On the other hand, if $\tau_{k_j}' \geq T$,

there arises a contradiction, because $\alpha < |x(t, \sigma_{k_j}, y_{k_j})|$ for

$\sigma_{k_j} < t \leq \tau'_{k_j}$, but $|x(T, \sigma_{k_j}, y_{k_j})| < B < \alpha$ for large j. Thus we see that the solutions of (8.2) are equi-bounded, which implies uniform boundedness since the system is periodic.

The second part of the conclusion follows from the following result for general systems.

Theorem 8.6. Consider the system (8.1) and assume that the solution of (8.1) is unique for the initial value problem. If the solutions of (8.1) are uniformly bounded and ultimately bounded, they are equi-ultimately bounded.

Proof. Let B be the bound for the ultimate boundedness, that is, there are $B > 0$ and $T = T(t_0, x_0)$ such that $|x(t, t_0, x_0)| < B$ for all $t \geq t_0 + T$. Let $\bar{B}$ be the bound for B by the uniform boundedness, that is, if $|x_0| \leq B$, then $|x(t, t_0, x_0)| < \bar{B}$ for all $t \geq t_0$.

Since $|x(t_0 + T, t_0, x_0)| = B' < B$, we take a neighborhood $\cup(P)$ of the point $P = x(t_0 + T, t_0, x_0)$ such that $\cup(P) \subset S = \{x; |x| \leq B\}$. By the uniqueness of solutions, there is a neighborhood $\cup^*(x_0)$ of the point x_0 such that $x(t_0 + T, t_0, x^*) \varepsilon \cup(P)$ if $x^* \varepsilon \cup^*(x_0)$. Then we have $|x(t, t_0, x^*)| < \bar{B}$ for all $t \geq t_0 + T$. Let S_α be the set of x such that $|x| \leq \alpha$. For each point $x \varepsilon S_\alpha$, consider such a neighborhood $\cup^*_x$ as the mentioned above. Since S_α is compact, it is covered by a finite number of $\cup^*$, say $\cup^*_i, 1 \leq i \leq k$. If we set $T = \max_i T_i$, where $T_i, 1 \leq i \leq k$, are determined by $\cup^*_i$, T depends only on t_0 and α. Then if $x_0 \varepsilon S_\alpha$, $|x(t, t_0, x_0)| < \bar{B}$ for all $t \geq t_0 + T$. This proves the theorem.

Now we shall discuss the boundedness of solutions of (8.1) by using Liapunov functions.

Theorem 8.7. Suppose that there exists a Liapunov function

$V(t,x)$ defined on $I \times R^n$ which satisfies the following conditions;

  (i) $a(|x|) \le V(t,x)$, where $a(r)$ is continuous and $a(r) \to \infty$

       as $r \to \infty$,

 (ii) $\dot{V}_{(8.1)}(t,x) \le 0$.

Then the solutions of (8.1) are equi-bounded.

  Theorem 8.8. Suppose that there exists a Liapunov function

$V(t,x)$ defined on $0 \le t < \infty$, $|x| \ge K$, where K can be large, which

satisfies the following conditions:

  (i) $a(|x|) \le V(t,x) \le b(|x|)$, where $a(r)$, $b(r)$ are continu-

       ous and $a(r) \to \infty$ as $r \to \infty$,

 (ii) $\dot{V}_{(8.1)}(t,x) \le 0$.

Then the solutions of (8.1) are uniformly bounded.

  For the proofs of these theorems, see [80].

  Example 8.3 [2]. Consider the equation

$$x'' + f(x,x')x' + g(x) = p(t), \qquad\qquad (8.5)$$

where we assume that

  (a) $f(x,y)$, $g(x)$ are continuous for all values of their

       variables,

  (b) $p(t)$ is continuous on I and $\int_0^\infty |p(t)|dt < \infty$,

  (c) $f(x,y) \ge 0$ for all x,y,

  (d) $G(x) = \int_0^x g(u)du > 0$ for all $x \ne 0$ and $G(x) \to \infty$

       as $|x| \to \infty$.

Then every solution $x(t)$ of (8.5) satisfies $|x(t)| < c$,

$|x'(t)| < c$, where c may depend on the solution.

  To see this, consider an equivalent system to (8.5)

$$x' = y, \quad y' = -f(x,y)y - g(x) + p(t) \tag{8.6}$$

and a Liapunov function $\quad V(t,x,y) = \sqrt{y^2+2G(x)} - \int_0^t |p(s)|ds \quad$ for

$x^2 + y^2 \geq K^2$. Since

$$\dot{V}_{(8.6)}(t,x,y) = \frac{1}{\sqrt{y^2+2G(x)}}\{g(x)y+y(-f(x,y)y-g(x)+p(t))-|p(t)|\}$$

$$\leq 0,$$

we can see that $V(t,x,y)$ satisfies all conditions in Theorem 8.8. Therefore the solutions of (8.6) are uniformly bounded, and thus we have $|x(t)| < c, \; |x'(t)| < c$.

Example 8.4. Suppose that $f(t,x)$ satisfies

$$|f(t,x)| \leq \lambda(t)\phi(|x|)$$

for $|x| \geq K$, where $\lambda(t)$ is continuous on I, $\int_0^\infty \lambda(t)dt < \infty$, $\phi(u)$ is continuous on $K \leq u < \infty$ and $\int_K^\infty \frac{du}{\phi(u)} = \infty$. Then the solutions of (8.1) are uniformly bounded, since the Liapunov function

$$V(t,x) = -\int_0^t \lambda(s)ds + \int_K^r \frac{du}{\phi(u)} , \quad r = |x| ,$$

satisfies the conditions in Theorem 8.8.

In some cases, the following theorem on boundedness is more convenient to apply. We consider a system

$$\begin{cases} x' = f(t,x,y) \\ y' = g(t,x,y), \end{cases} \tag{8.7}$$

where $f(t,x,y) \in C(I \times R^n \times R^m, R^n)$ and $g(t,x,y) \in C(I \times R^n \times R^m, R^m)$.

Theorem 8.9. Suppose that there exists a Liapunov function $V(t,x,y)$ defined on $0 \leq t < \infty$, $|x| + |y| \geq K$, where K can be large, which satisfies the following conditions;

> (i) $V(t,x,y)$ tends to infinity uniformly for (t,x) as
>
> $|y| \to \infty$,
>
> (ii) $V(t,x,y) \le b(|x|,|y|)$, where $b(r,s)$ is continuous,
>
> (iii) $\dot{V}_{(8.7)}(t,x,y) \le 0$.

Moreover, suppose that corresponding to each $M > 0$, there exists a

Liapunov function $W(t,x,y)$ defined on $0 \le t < \infty$, $|x| \ge K_1(M)$,

$|y| \le M$, where K_1 can be large, which satisfies the following condi-

tions;

> (iv) $W(t,x,y)$ tends to infinity uniformly for (t,y) as
>
> $|x| \to \infty$,
>
> (v) $W(t,x,y) \le c(|x|)$, where $c(r)$ is continuous,
>
> (vi) $\dot{W}_{(8.7)}(t,x,y) \le 0$.

Then the solutions of (8.7) are uniformly bounded.

   <u>Proof.</u> Let $x(t) = x(t,t_0,x_0,y_0)$, $y(t) = y(t,t_0,x_0,y_0)$ be

a solution of (8.7) such that $|x_0| + |y_0| \le \alpha$, $\alpha > K$. Choose a

$\beta(\alpha) > 0$ so large that

$$\sup_{\substack{|x|+|y|=\alpha \\ t \in I}} V(t,x,y) < \inf_{\substack{|y|=\beta \\ t \in I}} V(t,x,y).$$

This is possible by (i) and (ii). Then it can be seen that

$|y(t)| < \beta(\alpha)$ for $t \ge t_0$ as long as the solution exists. Now con-

sider $W(t,x,y)$ defined on $0 \le t < \infty$, $|x| \ge K_1(\beta)$, $|y| \le \beta$. Let α^*

be $\max\{\alpha, K_1(\beta)\}$, where α^* can be assumed to depend only on α.

Choose $\gamma(\alpha)$ so large that

$$\sup\{W(t,x,y); \ t \in I, \ |x| = \alpha^*, \ |y| \le \beta\}$$
$$< \inf\{W(t,x,y); \ t \in I, \ |x| = \gamma, \ |y| \le \beta\}.$$

Then, by (vi), it can be shown that $|x(t)| < \gamma(\alpha)$ as long as the

solution exists. Thus we can conclude that the solution $x(t), y(t)$

exists for all $t \geq t_0$ and that $|x(t)| < \gamma(\alpha)$ and $|y(t)| < \beta(\alpha)$

for all $t \geq t_0$. This proves that the solutions of (8.7) are uniformly

bounded.

Example 8.5. In the equation

$$x'' + f(x)x' + g(x) = p(t),   (8.8)$$

suppose that

(a) $f(x)$ is continuous and $F(x) = \int_0^x f(u)du \to \pm\infty$ as $x \to \pm\infty$,
respectively,

(b) $g(x)$ is continuous and $xg(x) \geq 0$ for $|x| > q$,

(c) $p(t)$ is continuous and $P(t) = \int_0^t p(s)ds$ is bounded.

Then every solution of (8.8) is bounded with its derivative.

To see this, consider a system equivalent to (8.8)

$$x' = y-F(x) + P(t),   y' = -g(x),   (8.9)$$

and choosing positive constants a,b suitably, define a Liapunov

function $V(t,x,y)$ in the following way: Setting $G(x) = \int_0^x g(u)du$

and $v(x,y) = G(x) + y^2/2$,

$$V(t,x,y) = \begin{cases} v(x,y) & (x \geq a,\ |y| < \infty) \\ v(x,y) - x+a & (|x| \leq a,\ y \geq b) \\ v(x,y) + 2a & (x \leq -a,\ y \geq b) \\ v(x,y) + \dfrac{2a}{b}\,y & (x \leq -a,\ |y| \leq b) \\ v(x,y) - 2a & (x \leq -a,\ y \leq -b) \\ v(x,y) + x-a & (|x| \leq a,\ y \leq -b). \end{cases}$$

Then we can see easily that $V(t,x,y)$ satisfies the conditions in

Theorem 8.9. Moreover, $W(t,x,y) = |x|$ defined for $|x| > K_1(M)$,

$|y| \leq M$ for suitable K_1 satisfies the conditions in Theorem 8.9. Therefore $x(t)$ and $y(t)$ of (8.9) are bounded and consequently $x'(t)$ also is bounded, because $y(t), F(x(t))$ and $P(t)$ are bounded.

Theorem 8.10. Under the assumptions in Theorem 8.8, if $\dot{V}_{(8.1)}(t,x) \leq -c(|x|)$, where $c(r)$ is positive and continuous, then the solutions of (8.1) are uniformly ultimately bounded.

Corollary 8.1. Under the assumptions in Theorem 8.8, if $\dot{V}_{(8.1)}(t,x) \leq -cV(t,x)$, where $c > 0$ is a constant, then the solutions of (8.1) are uniformly ultimately bounded.

Example 8.6. In Example 8.5, if $xg(x) > 0$ for $|x| \geq q > 0$ and $G(x) = \int_0^x g(u)du \to \infty$ as $|x| \to \infty$, we can see that the Liapunov function $V(t,x,y)$ defined in Example 8.5 satisfies the conditions in Theorem 8.10. Therefore the solutions of (8.9) are uniformly ultimately bounded, and thus we can find $B > 0$ independent of the solutions, for which $|x(t)| < B$, $|x'(t)| < B$ for large t.

Theorem 8.11. For the system (8.7), assume that there exists a Liapunov function $V(t,x,y)$ defined on $0 \leq t < \infty$, $|x| < \infty$, $|y| \geq K > 0$, which satisfies the following conditions;

   (i) $a(|y|) \leq V(t,x,y) \leq b(|y|)$, where $a(r)$ and $b(r)$ are continuous, increasing and $a(r) \to \infty$ as $r \to \infty$,

   (ii) $\dot{V}_{(8.7)}(t,x,y) \leq -c(|y|)$, where $c(r) > 0$ is continuous.

Suppose that corresponding to each M, there exists a Liapunov function $W(t,x,y)$ defined on $0 \leq t < \infty$, $|x| \geq K_1(M)$, $|y| \leq M$, which satisfies the following conditions;

   (iii) $a_1(|x|) \leq W(t,x,y) \leq b_1(|x|)$, where $a_1(r)$ and $b_1(r)$ are continuous and $a_1(r) \to \infty$ as $r \to \infty$,

(iv) $\dot{W}_{(8.7)}(t,x,y) \leq 0$.

Moreover, assume that letting B be such that $b(K) < a(B)$, there exists a Liapunov function $U(t,x,y)$ defined on $T \leq t < \infty$, $|x| \geq K_2 > 0$, $|y| \leq B$, which satisfies the following conditions;

    (v) $a_2(|x|) \leq U(t,x,y) \leq b_2(|x|)$, where $a_2(r)$ and $b_2(r)$ are continuous and increasing,

    (vi) $\dot{U}_{(8.7)}(t,x,y) \leq -c_2(|x|)$, where $c_2(r) > 0$ is continuous.

Then the solutions of (8.7) are uniformly ultimately bounded.

    Proof.   For an α such that $K < \alpha$, consider a solution $\{x(t) = x(t,t_0,x_0,y_0),\ y(t) = y(t,t_0,x_0,y_0)\}$ of (8.7), where $t_0 \geq 0$, $|x_0| \leq \alpha$ and $|y_0| \leq \alpha$. Choose a $\beta(\alpha) > 0$ so large that $b(\alpha) < a(\beta)$. We now show that $|y(t)| < \beta(\alpha)$ for $t \geq t_0$ as long as the solution $\{x(t),y(t)\}$ exists. Suppose that $|y(t_1)| = \beta(\alpha)$ at some t_1. Then there exist t_2 and t_3, $t_0 \leq t_2 < t_3 \leq t_1$, such that $|y(t_2)| = \alpha$, $|y(t_3)| = \beta$ and $\alpha < |y(t)| < \beta$ for $t_2 < t < t_3$. Consider the function $V(t,x(t),y(t))$ on $t_2 \leq t \leq t_3$. Then we have

$$a(\beta) \leq V(t_3,x(t_3),y(t_3)) \leq V(t_2,x(t_2),y(t_2)) \leq b(\alpha),$$

which contradicts $a(\beta) > b(\alpha)$. Therefore $|y(t)| < \beta(\alpha)$ for $t \geq t_0$ as long as the solution exists.

    Let $\alpha_1(\alpha) = \max\{\alpha, K_1(\beta(\alpha))\}$ and consider the Liapunov function $W(t,x,y)$ defined on $0 \leq t < \infty$, $|x| \geq K_1(\beta)$, $|y| \leq \beta$. Choose $\beta_1(\alpha)$ so large that $b_1(\alpha_1) < a_1(\beta_1)$ and suppose that $|x(t_1)| = \beta_1$ at some t_1. Then there exist t_2 and t_3, $t_0 \leq t_2 < t_3 \leq t_1$, such that $|x(t_2)| = \alpha_1$, $|x(t_3)| = \beta_1$ and $\alpha_1 < |x(t)| < \beta_1$, $|y(t)| < \beta(\alpha)$ for $t_2 \leq t \leq t_3$. Consider the function $W(t,x(t),y(t))$ on $t_2 \leq t \leq t_3$. Then

$$a_1(\beta_1) \leq W(t_3, x(t_3), y(t_3)) \leq W(t_2, x(t_2), y(t_2)) \leq b_1(\alpha_1),$$

which contradicts $a_1(\beta_1) > b_1(\alpha_1)$. Therefore, as long as the solution exists, $|x(t)| < \beta_1(\alpha)$ and $|y(t)| < \beta(\alpha)$ for $t \geq t_0$, which implies that the solution $\{x(t), y(t)\}$ exists for all $t \geq t_0$ and $|x(t)| < \beta_1(\alpha)$, $|y(t)| < \beta(\alpha)$ for all $t \geq t_0$. This means that the solutions of (8.7) are uniformly bounded.

Suppose now that $|y(t)| \geq K$ for all $t \geq t_0$. There is a $\lambda(\alpha) > 0$ such that if $|x| < \infty$ and $K \leq |y| \leq \beta(\alpha)$,
$\overset{\cdot}{V}_{(8.7)}(t,x,y) \leq -\lambda(\alpha)$. Therefore we have

$$V(t, x(t), y(t)) - V(t_0, x_0, y_0) \leq -\lambda(\alpha)(t-t_0).$$

If $t > t_0 + T_1(\alpha)$, where $T_1(\alpha) = \dfrac{b(\alpha) - a(K)}{\lambda(\alpha)}$,

$$a(K) \leq V(t, x(t), y(t)) \leq V(t_0, x_0, y_0) - \lambda(\alpha)(t-t_0)$$

$$< b(\alpha) - \lambda(\alpha)\dfrac{b(\alpha) - a(K)}{\lambda(\alpha)}$$

$$< a(K),$$

and hence there arises a contradiction. Therefore $|y(t_1)| < K$ at some t_1 such that $t_0 \leq t_1 \leq t_0 + T_1(\alpha)$. By the choice of B, we can see that $|y(t)| < B$ for all $t \geq t_1$. Thus we have $|y(t)| < B$ for all $t \geq t_0 + T_1(\alpha)$.

Now let $K_2^* = \max(B, K_2)$. Then, if $|x_0^*| \leq K_2^*$ and $|y_0^*| \leq K_2^*$,

$$|x(t, t_0, x_0^*, y_0^*)| < \beta_1(K_2^*) \quad \text{for all } t \geq t_0. \tag{8.10}$$

As was seen above, $|x(t)| < \beta_1(\alpha)$ for all $t \geq t_0$ and $|y(t)| < B$ for all $t \geq t_0 + T_1(\alpha)$, and consequently $|y(t)| < B$ for all $t \geq t_0 + T_1(\alpha) + T$. Suppose that $|x(t)| \geq K_2^*$ for all $t \geq t_0 + T_1(\alpha) + T$. For $K_2^* \leq |x| \leq \beta_1(\alpha)$, $|y| \leq B$ and $t \geq T$, there exists a $\lambda^*(\alpha) > 0$ such that

$$\dot{U}_{(8.7)}(t,x,y) \leq -\lambda^*(\alpha),$$

and hence

$$U(t,x(t),y(t)) \leq U(t_0+T_1(\alpha)+T,x(t_0+T_1(\alpha)+T),y(t_0+T_1(\alpha)+T))$$

$$- \lambda^*(\alpha)(t-t_0-T_1(\alpha)-T).$$

If $t > t_0 + T_1(\alpha) + T + T_2(\alpha)$, where $T_2(\alpha) = \dfrac{b_2(\beta_1(\alpha))-a_2(K_2^*)}{\lambda^*(\alpha)}$,

we have a contradiction. Therefore we have $|x(t_1)| < K_2^*$ at some t_1 such that $t_0+T_1(\alpha) + T \leq t_1 \leq t_0+T_1(\alpha)+T+T_2(\alpha)$, and hence, by (8.10), $|x(t)| < \beta_1(K_2^*)$ for all $t \geq t_1$, because $|x(t_1)| < K_2^*$ and $|y(t_1)| < B \leq K_2^*$. Thus we have

$$|x(t)| < \beta_1(K_2^*) \quad \text{for all} \quad t \geq t_0+T_1(\alpha)+T+T_2(\alpha)$$

and

$$|y(t)| < B \quad \text{for all} \quad t \geq t_0 + T_1(\alpha).$$

If we set $B^* = \beta_1(K_2^*)$ and $T^*(\alpha) = T_1(\alpha) + T + T_2(\alpha)$, clearly $B \leq B^*$ and $T_1(\alpha) \leq T^*(\alpha)$. Hence, if $|x_0| \leq \alpha$, $|y_0| \leq \alpha$ and $t_0 \geq 0$, we have

$$|x(t,t_0,x_0,y_0)| < B^*, \quad |y(t,t_0,x_0,y_0)| < B^*$$

for all $t \geq t_0+T^*(\alpha)$, where B^* is clearly a positive constant independent of particular solutions. This proves that the solutions of (8.7) are uniformly ultimately bounded.

Example 8.7. [60]. Consider an equation of third order

$$x''' + \phi(x')x'' + bx' + f(x) = p(t), \tag{8.11}$$

where $b > 0$ is a constant, $\phi(y)$, $f(x)$ and $p(t)$ are continuous. Consider an equivalent system

$$x' = y, \quad y' = z-\phi(y)+P(t), \quad z' = -f(x)-by, \tag{8.12}$$

where $\Phi(y) = \int_0^y \phi(u)\,du$ and $P(t) = \int_0^t p(s)\,ds$. The following assumptions will be made:

(a) $|f(x)| \le F$ for all x, and $f(x)\,\text{sgn}\,x > 0$ for $|x| \ge h$,

(b) $\phi(y)\,\text{sgn}\,y > 0$, for $|y| \ge k$, and $|\Phi(y)| \to \infty$ as $|y| \to \infty$,

(c) $|P(t)| \le m$ for all $t \ge 0$.

We shall show that the solutions of (8.12) are uniformly ultimately bounded. Let $K > 0$ be a constant such that

$$K \ge 1 + 2m + \frac{\alpha}{F} + 4F(1+b+\frac{1}{b}) + k,$$

where $\alpha = \max\{b|y|m + 2F|\Phi(y)| - by\Phi(y)\} \ge 0$, and that

$$|\Phi(y)| \ge 2m + \frac{F}{b}(1 + 2b + 2F + 4m)$$

for $|y| \ge K$. On the domain $0 \le t < \infty$, $\max(|y|-K, |z|-K) \ge 0$, consider a Liapunov function

$$V(y,z) = v(y,z) + u(y,z),$$

where $v(y,z) = \frac{1}{2}(by^2 + z^2)$ and

$$u(y,z) = \begin{cases} -2Fy\,\text{sgn}\,z & \text{for } |y| \le |z| \\ -2Fz\,\text{sgn}\,y & \text{for } |y| \ge |z| \end{cases}.$$

Clearly, $V(y,z)$ is continuous, positive and $V(y,z) \to \infty$ as $y^2+z^2 \to \infty$, because

$$\frac{1}{2}(by^2+z^2) - 2F|y| \ge \frac{1}{2}by^2 + \frac{1}{2}|z|(|z|-4F) > \frac{1}{2}by^2 \quad \text{for } |y| \le |z|,$$

$$\frac{1}{2}(by^2+z^2) - 2F|z| \ge \frac{1}{2}z^2 + \frac{1}{2}|y|(b|y|-4F) > \frac{1}{2}z^2 \quad \text{for } |y| \ge |z|,$$

and hence there exist continuous functions $a(r)$ and $b(r)$ such that

$$a(|y|+|z|) \le V(y,z) \le b(|y|+|z|).$$

On the other hand, $\dot{v}_{(8.12)}(y,z) \leq -by\Phi(y)+b|y|m + |z|F$ and

$$\dot{u}_{(8.12)}(y,z) = -2F|z| + 2F\Phi(y)\,\text{sgn } z - 2FP(t)\,\text{sgn } z$$
$$\leq -2F|z| + 2F|\Phi(y)| + 2Fm$$

for $|y| \leq |z|$ and

$$\dot{u}_{(8.12)}(y,z) = 2Ff(x)\,\text{sgn } y + 2Fb|y| \leq 2F^2 + 2Fb|y|$$

for $|y| \geq |z|$. Therefore, for $|y| \geq |z|$,

$$\dot{v}_{(8.12)}(y,z) \leq -|y|\{b|\Phi(y)|-bm-F-2Fb-2F^2\} < 0,$$

because $|y| \geq K$, and for $|y| \leq |z|$, $|y| \geq K$

$$\dot{v}_{(8.12)}(y,z) \leq -\frac{b}{2}|y|\{|\Phi(y)|-2m\}-\frac{1}{2}|\Phi(y)|\{b|y|-4F\}-F(|y|-2m) < 0,$$

and for $|y| \leq |z|$, $|y| \leq K$

$$\dot{v}_{(8.12)}(y,z) \leq 2Fm + \alpha-F|z| < 0 \quad (\text{by } |z| \geq K).$$

Now consider a function $W(x,z) = \frac{b}{2}(x + \frac{z}{b})^2$ on $|z| \leq M$
and $|x| \geq \max(h, \frac{M}{b})$. Then, clearly

$$\dot{W}_{(8.12)}(x,z) = -(x + \frac{z}{b})f(x) < 0.$$

Letting B be such that $b(K) < a(B)$, consider $U(x,z) = \frac{b}{2}(x+ \frac{z}{b})^2$
on $|z| \leq B$, $|x| \geq \max(h, \frac{2B}{b})$. Then we have

$$\frac{b}{8}x^2 \leq U(x,z) \leq \frac{9b}{8}x^2$$

and

$$\dot{U}_{(8.12)}(x,z) \leq -\frac{1}{2}xf(x) < 0.$$

Thus, applying Theorem 8.11, we can see that the solutions of (8.12)
are uniformly ultimately bounded.

9. Asymptotic Stability in the Large

Consider a system

$$x' = f(t,x), \quad f(t,0) \equiv 0, \qquad (9.1)$$

where $f(t,x) \in C(I \times R^n, R^n)$.

Definition 9.1. The zero solution of (9.1) is <u>asymptotically stable in the large</u>, if it is stable and if every solution of (9.1) tends to zero as $t \to \infty$.

Definition 9.2. The zero solution of (9.1) is <u>quasi-equi-asymptotically stable in the large</u>, if for any $\alpha > 0$, any $\varepsilon > 0$ and $t_0 \in I$, there exists a $T(t_0, \varepsilon, \alpha) > 0$ such that if $|x_0| \leq \alpha$, then $|x(t, t_0, x_0)| < \varepsilon$ for all $t \geq t_0 + T(t_0, \varepsilon, \alpha)$.

Definition 9.3. The zero solution of (9.1) is <u>equi-asymptotically stable in the large</u>, if it is stable and is quasi-equiasymptotically stable in the large.

Definition 9.4. The zero solution of (9.1) is <u>quasi-uniformly asymptotically stable in the large</u>, if the T in Definition 9.2 is independent of t_0.

Definition 9.5. The zero solution of (9.1) is <u>uniformly asymptotically stable in the large</u>, if it is uniformly stable and is quasi uniformly asymptotically stable in the large and if the solutions of (9.1) are uniformly bounded.

Definition 9.6. The zero solution of (9.1) is <u>exponentially asymptotically stable in the large</u>, if there exists a $c > 0$ and for any $\alpha > 0$, there exists a $K(\alpha) > 0$ such that if $|x_0| \leq \alpha$,

$$|x(t, t_0, x_0)| \leq K(\alpha) e^{-c(t-t_0)} |x_0| \quad \text{for all } t \geq t_0. \qquad (9.2)$$

If x(t) ≡ 0 is the unique solution of (9.1) through (0,0),
quasi-equiasymptotic stability in the large implies equiasymptotic
stability in the large. As Example 6.1 shows, quasi uniformly asymptotic stability in the large does not necessarily imply uniformly
asymptotic stability in the large. For the linear system (7.2), the
following properties can be easily seen.

Theorem 9.1. If the zero solution of (7.2) is asymptotically
stable, it is asymptotically stable in the large. Moreover, if the
zero solution of (7.2) is uniformly asymptotically stable, it is exponentially asymptotically stable in the large, and in this case, we
can find a $K > 0$ independent of α in (9.2).

Theorem 9.2. For the linear system (7.2),

(a) asymptotic stability and ultimate boundedness are equi-
     valent, and consequently equiasymptotic stability and
     equiultimate boundedness are equivalent,

(b) quasi uniformly asymptotic stability in the large and
     uniformly ultimate boundedness are equivalent.

Theorem 9.3. If the zero solution of the periodic system (8.2)
is asymptotically stable in the large, then it is uniformly asymptotically stable in the large.

We shall now consider a system

$$x' = f(t,x), \tag{9.3}$$

where $f(t,x) \in C(I \times R^n, R^n)$. We assume that (9.3) has a solution
$\phi(t)$ defined on I.

Definition 9.7. The solution $\phi(t)$ is said to be weakly uniformly asymptotically stable in the large, if it is uniformly stable

and for every $t_0 \in I$ and every $x_0 \in R^n$, we have

$|x(t,t_0,x_0)-\phi(t)| \to 0$ as $t \to \infty$.

The following theorem can be proved by the same argument as in the proof of Theorem 7.6 [82].

Theorem 9.4. If the solution $\phi(t)$ is weakly uniformly asymptotically stable in the large, then it is equiasymptotically stable in the large. Namely for any $\varepsilon > 0$, any $\alpha > 0$ and $t_0 \in I$ there exists a $T(t_0,\varepsilon,\alpha) \geq 0$ such that $|x_0-\phi(t_0)| \leq \alpha$ implies that

$$|x(t,t_0,x_0)-\phi(t)| < \varepsilon \text{ for all } t \geq t_0 + T(t_0,\varepsilon,\alpha).$$

Theorem 9.5. Suppose that the system (9.3) is periodic in t of period ω and has a bounded solution $\phi(t)$ which is weakly uniformly asymptotically stable in the large. Then the solution $\phi(t)$ is uniformly asymptotically stable in the large.

Proof. Let $B > 0$ be such that $|\phi(t)| \leq B$ for all $t \geq 0$. Since $|\phi(t)| \leq B$ and $\phi(t)$ is equiasymptotically stable in the large by Theorem 9.4, we can see that the solutions of (9.3) are equibounded. Therefore, for any $\alpha > 0$ and $t_0 \in I$, we can find a $\beta(\alpha) > 0$ such that if $t_0 \in I$ and $|x_0-\phi(t_0)| \leq \alpha$, then $|x(t,t_0,x_0)-\phi(t)| < \beta(\alpha)$ for all $t \geq t_0$.

By the assumption, $\phi(t)$ is uniformly stable, and hence it is sufficient to show that for any $\varepsilon > 0$ and $\alpha > 0$, there exists a $T(\varepsilon,\alpha) > 0$ such that if $|x_0-\phi(t_0)| \leq \alpha$, then

$$|x(t,t_0,x_0)-\phi(t)| < \varepsilon \text{ for all } t \geq t_0 + T(\varepsilon,\alpha).$$

Given $\alpha > 0$, if $0 \leq t_0 < \omega$ and $|x_0-\phi(t_0)| \leq 2B + \alpha$, then

$$|x(\omega,t_0,x_0)-\phi(\omega)| < \beta(2B + \alpha).$$

By equiasymptotic stability in the large, there exists a $T_1(\omega, \frac{\varepsilon}{2}, \alpha) > 0$ such that if $|x_0 - \phi(\omega)| < \beta(2B+\alpha)$, then $|x(t,\omega,x_0) - \phi(t)| < \frac{\varepsilon}{2}$ for all $t \geq \omega + T_1(\omega, \frac{\varepsilon}{2}, \alpha)$.

Now consider a solution $x(t,t_0,x_0)$ such that $|x_0 - \phi(t_0)| \leq \alpha$ and $k\omega \leq t_0 < (k+1)\omega$, where $k = 0,1,2,\ldots$. Since the system (9.3) is periodic in t of period ω, we have

$$x(t,t_0,x_0) = x(t-k\omega,t_0-k\omega,x_0), \quad t \geq t_0. \qquad (9.4)$$

Moreover, $\phi(t+k\omega)$ also is a solution of (9.3) such that $x = \phi(k\omega)$ at $t = 0$, which we shall denote by $\psi(t,0,\phi(k\omega))$. Then we have

$$|\psi(\omega,0,\phi(k\omega)) - \phi(\omega)| \leq 2B \quad (< \beta(2B+\alpha)),$$

and hence we have

$$|\psi(t,0,\phi(k\omega)) - \phi(t)| < \frac{\varepsilon}{2} \quad \text{for all} \quad t \geq \omega + T_1(\omega, \frac{\varepsilon}{2}, \alpha). \qquad (9.5)$$

Since $|x(t_0,t_0,x_0) - \phi(t_0)| \leq \alpha$ and $\phi(t_0) = \psi(t_0-k\omega,0,\phi(k\omega))$, it follows from (9.4) that

$$|x(t_0-k\omega,t_0-k\omega,x_0) - \psi(t_0-k\omega,0,\phi(k\omega))| \leq \alpha,$$

which implies that $|x(t_0-k\omega,t_0-k\omega,x_0) - \phi(t_0-k\omega)| \leq 2B+\alpha$, because $|\psi(t_0-k\omega,0,\phi(k\omega)) - \phi(t_0-k\omega)| \leq 2B$. Therefore we have

$$|x(t-k\omega,t_0-k\omega,x_0) - \phi(t-k\omega)| < \frac{\varepsilon}{2} \qquad (9.6)$$

for all $t \geq (k+1)\omega + T_1(\omega, \frac{\varepsilon}{2}, \alpha)$ since $0 \leq t_0-k\omega < \omega$. From (9.5) it follows that

$$|\psi(t-k\omega,0,\phi(k\omega)) - \phi(t-k\omega)| < \frac{\varepsilon}{2} \qquad (9.7)$$

for $t \geq (k+1)\omega + T_1(\omega, \frac{\varepsilon}{2}, \alpha)$. Thus, by (9.6) and (9.7)

$$|x(t-k\omega,t_0-k\omega,x_0) - \psi(t-k\omega,0,\phi(k\omega))| < \varepsilon$$

for all $t \geq (k+1)\omega + T_1(\omega,\frac{\varepsilon}{2},\alpha)$, which implies that

$$|x(t,t_0,x_0) - \phi(t)| < \varepsilon \quad \text{for all} \quad t \geq t_0 + T(\varepsilon,\alpha),$$

where $T(\varepsilon,\alpha) = \omega + T_1(\omega,\frac{\varepsilon}{2},\alpha)$, because $t_0 \geq k\omega$. Thus we see that the solution $\phi(t)$ is uniformly asymptotically stable in the large.

The following example is of a scalar almost periodic equation such that the zero solution is weakly uniformly asymptotically stable in the large but is not uniformly asymptotically stable in the large [62].

Example 9.1. Consider the equation

$$x' = g(t,x), \qquad\qquad (9.8)$$

where

$$g(t,x) = \begin{cases} -x & (0 \leq x \leq 1) \\ -1 + (1-2f(t))(x-1) & (1 < x \leq 2) \\ -f(t)x & (2 < x) \end{cases}$$

and $g(t,x) = -g(t,-x)$ for $x < 0$. Here $f(t)$ is the almost periodic function constructed by Conley and Miller, which we have considered in equation (6.7).

The zero solution is clearly uniformly asymptotically stable. We can assume that $|f(t)| \leq 1$. Then

$$g(t,x) \leq -f(t)x.$$

Comparing with the solution of $x' = -f(t)x$, we can see that every solution of (9.8) tends to zero as $t \to \infty$. Thus the zero solution is weakly uniformly asymptotically stable in the large. As was seen in Section 6, if we consider the solution of (6.7) through (t_n,x_0), $x_0 > 2$, then

$$x(t_n + T_n,t_n,x_0) \geq x_0 e^n.$$

Therefore there exist t_n' and T_n' such that

$$t_n \leq t_n' < t_n' + T_n' \leq t_n + T_n,$$

$$x(t_n',t_n,x_0) = x_0,$$

$$x(t_n' + T_n',t_n,x_0) = x_0 e^n$$

and that

$$x_0 < x(t,t_n,x_0) < x_0 e^n \quad \text{for} \quad t_n' < t < t_n' + T_n'.$$

However the solution of (9.8) coincides with the solution of (6.7) on the interval $t_n' \leq t \leq t_n' + T_n'$. Thus the solutions of (9.8) are not uniformly bounded, and hence the zero solution of (9.8) is not uniformly asymptotically stable in the large.

Theorem 9.6. Suppose that there exists a Liapunov function $V(t,x)$ defined on $I \times R^n$ which satisfies the following conditions;

(i) $V(t,0) \equiv 0$,

(ii) $a(|x|) \leq V(t,x)$, where $a(r)$ is continuous, increasing, positive definite and $a(r) \to \infty$ as $r \to \infty$,

(iii) $\dot{V}_{(9.1)}(t,x) \leq -cV(t,x)$, where $c > 0$ is a constant.

Then the solution $x(t) \equiv 0$ of (9.1) is equiasymptotically stable in the large.

Proof. By Theorem 6.2, the zero solution is stable. Moreover, we can easily see that the solutions are equi-bounded and hence, every solution exists in the future. Let $x(t,t_0,x_0)$ be a solution such that $|x_0| \leq \alpha$. Applying Theorem 1.1, by (iii)

$$V(t,x(t,t_0,x_0)) \leq V(t_0,x_0)e^{-c(t-t_0)}. \qquad (9.9)$$

Let $M(t_0,\alpha) = \max\{V(t_0,x_0); |x_0| \leq \alpha\}$ and let $T(t_0,\varepsilon,\alpha)$ be such that

$$T(t_0, \varepsilon, \alpha) = \frac{1}{c} \log \frac{M(t_0, \alpha)}{a(\varepsilon)} \ .$$

Then it follows from (9.9) that

$$V(t, x(t, t_0, x_0)) < M(t_0, \alpha) \frac{a(\varepsilon)}{M(t_0, \alpha)} = a(\varepsilon)$$

$$\text{for} \quad t > t_0 + T(t_0, \varepsilon, \alpha) .$$

Since $a(r)$ is increasing and $a(|x|) \le V(t, x)$, we have

$|x(t, t_0, x_0)| < \varepsilon$ for $t > t_0 + T(t_0, \varepsilon, \alpha)$, which proves that $x(t) \equiv 0$

is quasi-equiasymptotically stable in the large. This completes the

proof.

The following is a sufficient condition for asymptotic stability

in the large.

Theorem 9.7. Suppose that there exists a Liapunov function

$V(t, x)$ defined on $I \times R^n$ which satisfies the following conditions;

> (i) $V(t, 0) \equiv 0$,
>
> (ii) $a(|x|) \le V(t, x)$, where $a(r)$ is continuous, positive
>
> definite and $a(r) \to \infty$ as $r \to \infty$.
>
> (iii) $\dot{V}_{(9.1)}(t, x) \le -c(x)$, where $c(x)$ is continuous and
>
> positive definite.

Then, if $f(t, x)$ is bounded when x belongs to a compact set or if

$$\overline{\lim_{h \to 0^+}} \frac{1}{h} \{ c(x(t+h)) - c(x(t)) \} \tag{9.10}$$

is bounded from above (or below), where $x(t)$ is any solution of

(9.1), then the zero solution is asymptotically stable in the large.

Proof. It is easily seen that the zero solution is stable and

the solutions of (9.1) are equi-bounded, that is, for any $\alpha > 0$ and

any $t_0 \in I$, there is a $\beta(t_0, \alpha) > 0$ such that $|x_0| \le \alpha$ implies

$|x(t,t_0,x_0)| < \beta(t_0,\alpha)$ for all $t \geq t_0$.   In the case where $f(t,x)$
is bounded when x belongs to a compact set, the asymptotic stability
in the large can be proved by the same argument in the proof of
Theorem 7.10.

Now consider a solution $x(t) = x(t,t_0,x_0)$ such that $|x_0| \leq \alpha$.
Then $|x(t)| < \beta(t_0,\alpha)$ for all $t \geq t_0$ and we suppose that (9.10)
for this solution is bounded by K from above.   Since

$$\dot{V}(t,x(t)) \leq -c(x(t)),$$

we have $\int_{t_0}^{t} c(x(s))ds \leq V(t_0,x_0) - V(t,x(t))$, but $c(x(t)) \geq 0$ and
$V(t,x(t)) \geq 0$, and hence

$$\int_{t_0}^{\infty} c(x(t))dt < \infty. \qquad (9.11)$$

Suppose that $c(x(t))$ does not tend to zero as $t \to \infty$.   Then,
for some $\varepsilon > 0$ there is a sequence $\{t_k\}$ such that $t_k \to \infty$ as
$k \to \infty$ and $c(x(t_k)) \geq \varepsilon$.   On the intervals

$$t_k - \frac{\varepsilon}{2K} \leq t \leq t_k, \qquad (9.12)$$

we have $c(x(t)) \geq \frac{\varepsilon}{2}$.   In the case where (9.10) is bounded by
$-K, K > 0$, from below, consider the intervals $t_k \leq t \leq t_k + \frac{\varepsilon}{2K}$.   We
can assume that these intervals are disjoint.   Thus we have

$$\int_{t_0}^{\infty} c(x(t))dt = \infty,$$

which contradicts (9.11).   Thus $c(x(t)) \to 0$ as $t \to \infty$.   Since $c(x)$
is continuous, positive definite and $|x(t)| \leq \beta(t_0,\alpha)$, $x(t) \to 0$ as
$t \to \infty$.   This proves that the zero solution is asymptotically stable
in the large.

Theorem 9.8.   Suppose that there exists a Liapunov function
$V(t,x)$ defined on $I \times R^n$ which satisfies the following conditions;

(i) $a(|x|) \leq V(t,x) \leq b(|x|)$, where $a(r)$ and $b(r)$ are continuous, positive definite and $a(r) \to \infty$ as $r \to \infty$,

(ii) $\dot{V}_{(9.1)}(t,x) \leq -c(|x|)$, where $c(r)$ is continuous and positive definite.

Then the zero solution of (9.1) is uniformly asymptotically stable in the large.

Let us consider <u>Liénard's equation</u>

$$x'' + f(x)x' + g(x) = 0, \tag{9.13}$$

where $f(x)$, $g(x)$ are continuous on $x \varepsilon R^1$. Suppose that $g(x)F(x) > 0$ for $x \neq 0$, where $F(x) = \int_0^x f(u)du$, and that $xg(x) > 0$

for $x \neq 0$ and $G(x) = \int_0^x g(u)du \to \infty$ as $|x| \to \infty$.

Consider an equivalent system

$$x' = y - F(x), \quad y' = -g(x) \tag{9.14}$$

and a Liapunov function $V(t,x,y) = G(x) + y^2/2$. Clearly $V(t,0,0) = 0$ and $V(t,x,y) \to \infty$ uniformly as $x^2+y^2 \to \infty$. Since we have

$$\dot{V}_{(9.14)}(t,x,y) = -g(x)F(x) \leq 0,$$

the zero solution of (9.14) is uniformly stable and the solutions of (9.14) are uniformly bounded. However, $\dot{V}_{(9.14)}(t,x,y)$ does not satisfy condition (ii) in Theorem 9.8, and hence Theorem 9.8 cannot be applied to this case. In fact, the zero solution of (9.14) is uniformly asymptotically stable in the large. For this reason we shall discuss some extensions of stability theory in the following section.

10. <u>Asymptotic Behavior of Solutions</u>

First of all, under the assumption that a solution $x(t)$ is bounded and approaches a closed set Ω, it will be shown that the

positive limit set of x(t) is composed of solutions of some system
defined on Ω which is related to the unperturbed system.

Lemma 10.1.   Consider a system

$$x' = f(t,x),   (10.1)$$

where f(t,x) is continuous on an open set D in R^{n+1}. Suppose
that every solution of (10.1) starting from a point (t_0,x_0) ε D to
the right is continuable to t = T. Let F denote the family of solutions of (10.1) on $J = [t_0,T]$ which pass through (t_0,x_0) and let
E be the set of all points (t,x) such that (t,x) is on some solution in F. Assume that E is contained in a compact set in D.
Then, corresponding to each ε > , there exists a $\delta > 0$ such that if
$d(P,E) \leq \delta$, $P = (\tau,\xi)$, τ ε J, then every solution $y(t,\tau,\xi)$ of

$$x' = f(t,x) + g(t),   (10.2)$$

where g(t) is a continuous function such that $\int_{t_0}^{T} |g(t)|\,dt \leq \delta$,
exists on $[\tau,T]$ and there exists a solution $x(t,t_0,x_0)$ of (10.1)
such that $(t,x(t,t_0,x_0))$ ε E and

$$|x(t,t_0,x_0)-y(t,\tau,\xi)| < \varepsilon \quad \text{for} \quad t \geq \tau,$$

where $x(t,t_0,x_0)$ may depend on $y(t,\tau,\xi)$.
    For the proof, see [76], [80].

Definition 10.1.   For a system defined on a set D

$$x' = f(x),   x \varepsilon D,   (10.3)$$

and for a subset M of D, M is said to be a semi-invariant set of
(10.3), if for each point of M there exists at least one solution
of (10.3) which remains in M for all future time.

    Consider a system

$$x' = f(t,x) + g(t,x). \tag{10.4}$$

Let Q be an open set in R^n and suppose that $f(t,x)$, $g(t,x)$ are continuous on $I \times Q$. Moreover, suppose that if $x(t)$ is continuous and bounded on $[t_0,\infty)$, that is, for some compact set $Q^* \subset Q, x(t) \in Q^*$ for all $t \in [t_0,\infty)$, then we have

$$\int_{t_0}^{\infty} |g(s,x(s))|ds < \infty. \tag{10.5}$$

Let $x(t,t_0,x_0)$ be a solution of (10.4) through (t_0,x_0) and let Γ^+ be a set in R^n such that

$$\Gamma^+ = \bigcap_{t_0 \leq s < \tau} \overline{\bigcup_{s \leq t} x(t,t_0,x_0)},$$

where the interval $[t_0,\tau)$ is the maximal interval of the solution $x(t,t_0,x_0)$. Then Γ^+ is a closed subset in Q, and if $x(t,t_0,x_0)$ is bounded, Γ^+ is non-empty and compact.

Lemma 10.2. Suppose that $f(t,x)$ and $g(t,x)$ are continuous on $I \times Q$. If Γ^+ is non-empty, then we have $\tau = +\infty$.

For the proof, see [80].

As is well known, Γ^+ is the <u>positive limit set</u> of the solution $x(t,t_0,x_0)$. If Γ^+ is non-empty, $x(t) = x(t,t_0,x_0)$ is defined on $[t_0,\infty)$ by Lemma 10.2. In this case, the existence of a point ω in Γ^+ is equivalent to saying that there exists a sequence $\{t_k\}$ such that $x(t_k) \to \omega$, $t_k \to \infty$ as $k \to \infty$.

Lemma 10.3. Let $x(t)$ be a solution of (10.4) bounded for $t \geq t_0$ and let Γ^+ be the positive limit set of $x(t)$. Then $x(t) \to \Gamma^+$ as $t \to \infty$. Moreover, if $x(t)$ is bounded for $t \geq t_0$ and if M contains Γ^+, then $x(t) \to M$ as $t \to \infty$.

Lemma 10.4. Let Ω be a closed set in the space Q. Suppose that a solution $x(t)$ approaches Ω as $t \to \infty$. Then the positive limit set Γ^+ of $x(t)$ is contained in Ω.

These lemmas can be easily verified.

Now we shall make the following assumption for the system (10.4). Let Ω be a nonempty closed set in the space Q and suppose that $f(t,x)$ of (10.4) satisfies the following conditions:

(a) $f(t,x)$ tends to a function $h(x)$ for $x \in \Omega$ as $t \to \infty$ and on any compact set in Ω this convergence is uniform.

(b) Corresponding to each $\varepsilon > 0$ and each $y \in \Omega$, there exists a $\delta(\varepsilon,y) > 0$ and a $T(\varepsilon,y) > 0$ such that if $|x-y| < \delta(\varepsilon,y)$ and $t \geq T(\varepsilon,y)$, we have $|f(t,x)-f(t,y)| < \varepsilon$.

Remark. Here, since $f(t,x)$ is defined on I, we can choose $\delta(\varepsilon,y)$ so that $|f(t,x)-f(t,y)| < \varepsilon$ for all $t \geq 0$, if we have condition (b). However, in case $f(t,x)$ is defined on $0 < t < \infty$, we require the existence of $T(\varepsilon,y)$ to obtain our results.

The following lemma can be proved in the same manner as in the standard proof of uniform continuity of a continuous function on a compact set.

Lemma 10.5. If $f(t,x)$ satisfies condition (b), for $y \in \Omega_1$, where Ω_1 is a compact set in Ω, we can choose δ and T which are independent of y and depend only on Ω_1.

Theorem 10.1. Suppose that a solution $x(t,t_0,x_0)$ of (10.4) is bounded and approaches a closed set Ω in the space Q. If $f(t,x)$ satisfies conditions (a) and (b), then the positive limit set Γ^+ of $x(t,t_0,x_0)$ is a semi-invariant set contained in Ω of the equation

$$x' = h(x), \quad x \in \Omega. \tag{10.6}$$

Proof. Since $x(t) = x(t,t_0,x_0)$ is bounded in Q, there exists
a compact set Q* in Q such that $x(t) \in Q*$ for all $t \geq t_0$. By
Lemma 10.4, $\Gamma^+ \subset \Omega \cap Q* = \Omega_1$. Since Ω_1 is a compact set in R^n,
there exists a continuous, bounded function $h*(x)$ on R^n such that
$h*(x) = h(x)$ on Ω_1. Let ω be a point of Γ^+. Then $\omega \in \Omega_1$ and
there exists a sequence $\{t_k\}$ such that

$$x(t_k) \rightarrow \omega, \; t_k \rightarrow \infty \;\; \text{as} \;\; k \rightarrow \infty. \tag{10.7}$$

Now consider the systems

$$x' = h*(x), \;\; x \in R^n \tag{10.8}$$

and

$$x' = h*(x)+f(t+t_k,x(t+t_k))-h*(x(t+t_k))+g(t+t_k,x(t+t_k)). \tag{10.9}$$

Since $h*(x)$ is bounded, for any $\lambda > 0$, all solutions of (10.8) are
defined on $0 \leq t \leq \lambda$. It is clear that $x(t+t_k)$ is a solution of
(10.9) through $(0,x(t_k))$. Since $x(t)$ is bounded, we have

$$\int_{t_0}^{\infty} |g(s,x(s))| ds < \infty$$

by condition (10.5), and hence, if k is sufficiently large, say
$k \geq k_1$, for a given $\delta > 0$

$$\int_{t_k}^{t_k+\lambda} |g(s,x(s))| ds < \frac{\delta}{2}$$

or

$$\int_0^{\lambda} |g(s+t_k,x(s+t_k))| ds < \frac{\delta}{2}. \tag{10.10}$$

For every point $x(t+t_k)$, there is a point $y(t+t_k) \in \Omega_1$ such that

$$d(x(t+t_k),\Omega_1) = |x(t+t_k) - y(t+t_k)|,$$

because Ω_1 is a compact set.

From condition (b) and Lemma 10.5, it follows that, correspond-

ing to $\frac{\delta}{6\lambda}$, there are $\delta_1 > 0$ and $T > 0$ such that $y \in \Omega_1$,
$|x-y| < \delta_1$ and $t \geq T$ imply $|f(t,x)-f(t,y)| < \frac{\delta}{6\lambda}$. On the other hand,
$x(t+t_k) \in N(\delta_1,\Omega_1) \cap Q^*$ for sufficiently large k, because $x(t) \to \Omega_1$
as $t \to \infty$. Therefore, if k is sufficiently large, say $k \geq k_2$, we
have

$$|f(t+t_k,x(t+t_k))-f(t+t_k,y(t+t_k))| < \frac{\delta}{6\lambda} \qquad (10.11)$$

on the interval $0 \leq t \leq \lambda$.

By condition (a), $f(t,x) \to h(x)$ uniformly in $x \in \Omega_1$ as
$t \to \infty$, and hence

$$|f(t,x)-h(x)| < \frac{\delta}{6\lambda}$$

for sufficiently large t and for $x \in \Omega_1$. Therefore, if $k \geq k_3$
for sufficiently large $k_3 > 0$ and t is in $[0,\lambda]$, we have

$$|f(t+t_k,y(t+t_k))-h(y(t+t_k))| < \frac{\delta}{6\lambda} . \qquad (10.12)$$

Moreover, there exists a $\delta_2 > 0$ such that $|x-y| < \delta_2$ implies
$|h^*(x)-h^*(y)| < \frac{\delta}{6\lambda}$ since $h^*(x)$ is continuous on Q^*. Thus, if k
is sufficiently large, say $k \geq k_4$, we have

$$|h^*(x(t+t_k))-h^*(y(t+t_k))| < \frac{\delta}{6\lambda} \qquad (10.13)$$

on $0 \leq t \leq \lambda$. Since $h^*(y(t+t_k)) = h(y(t+t_k))$, it follows from
(10.11), (10.12) and (10.13) that if $k > \max(k_1,k_2,k_3,k_4)$,

$$|f(t+t_k,x(t+t_k))-h^*(x(t+t_k))| < \frac{\delta}{2\lambda} , \quad 0 \leq t \leq \lambda. \qquad (10.14)$$

From (10.10) and (10.14), we have

$$\int_0^\lambda |f(s+t_k,x(s+t_k))-h^*(x(s+t_k))+g(s+t_k,x(s+t_k))|ds < \delta.$$

On the other hand, $|x(t_k)-\omega| < \delta$ for sufficiently large k
by (10.7). Therefore, applying Lemma 10.1, there exists a solution

$\phi_k(t)$ defined on $0 \leq t \leq \lambda$ of system (10.8) through $(0,\omega)$ such

that for a given $\varepsilon > 0$, $|x(t+t_k)-\phi_k(t)| < \varepsilon$ for $t \varepsilon [0,\lambda]$. Since

$x(t) \rightarrow \Gamma^+$ as $t \rightarrow \infty$, if k is sufficiently large, $\phi_k(t) \varepsilon N(2\varepsilon,\Gamma^+)$

for $t \varepsilon [0,\lambda]$, and also

$$\phi_k(t) = \omega + \int_0^t h^*(\phi_k(s))ds \quad \text{for} \quad t \varepsilon [0,\lambda].$$

Thus, for a sequence $\{\varepsilon_k\}$ approaching zero as $k \rightarrow \infty$, there exist

solutions $\phi_k(t)$ of (10.8) such that

$$\left\{ \begin{array}{l} \phi_k(t) = \omega + \int_0^t h^*(\phi_k(s))ds \\[3mm] \phi_k(t) \varepsilon N(\varepsilon_k,\Gamma^+) \end{array} \right. \tag{10.15}$$

for $t \varepsilon [0,\lambda]$. Since $\{\phi_k(t)\}$ is uniformly bounded and equicontin-

uous, it has a uniformly convergent subsequence. Let $\phi(t)$ be its

limit function. Then, by (10.15),

$$\phi(t) = \omega + \int_0^t h^*(\phi(s))ds, \phi(t) \varepsilon \Gamma^+ \quad \text{for} \quad t \varepsilon [0,\lambda].$$

Since $\Gamma^+ \subset \Omega_1$, $h^*(\phi(t)) = h(\phi(t))$, which implies that

$$\phi(t) = \omega + \int_0^t h(\phi(s))ds \quad \text{for} \quad t \varepsilon [0,\lambda],$$

that is, $\phi(t)$ is a solution of system (10.6) through $(0,\omega)$ and re-

mains in Γ^+. Since λ is arbitrary, we can find a solution of

(10.6) defined on I which passes through $(0,\omega)$ and remains in

Γ^+. This proves that Γ^+ is a semi-invariant set of (10.6).

Remark 1. As we can see from the proof above, in the case

where $f(t,x)$ and $g(t,x)$ are defined for $t \varepsilon (0,\infty)$, $x \varepsilon Q$, the

conclusion of Theorem 10.1 is also true for a solution $x(t,t_0,x_0)$

such that $t_0 > 0$ and $x_0 \varepsilon Q$.

Remark 2. If k is so large that $[t_k-\lambda,t_k]$ is contained in

I, the same argument can be applied to showing that there is a solution of (10.6) defined on $-\infty < t \leq 0$ which passes through $(0,\omega)$ and remains in Γ^+. Therefore Γ^+ is a <u>semi-invariant set in both directions</u>.

<u>Corollary 10.1.</u> Assume that $f(t,x)$ in (10.4) satisfies conditions (a) and (b) for a closed set Ω in the space Q. If for a solution $x(t)$ of (10.4) approaching Ω, we have $\lim_{t\to\infty} x(t) = x_0$, then the point x_0 is a critical point of (10.6), that is, $h(x_0) = 0$.

We stated Theorem 10.1 in a special form which is convenient for applications, but the proof of Theorem 10.1 is easily modified so as to be applied to more general case, see [75].

<u>Theorem 10.2.</u> Suppose that the positive limit set Γ^+ of $x(t,t_0,x_0)$ of (10.4) is non-empty and $x(t,t_0,x_0) \to \Omega$ as $t \to \infty$, where Ω is a closed set in the space Q, and suppose that $f(t,x)$ satisfies conditions (a) and (b). Then Γ^+ is the union of solutions of (10.6).

Using the results above and a <u>Liapunov function</u>, we shall obtain some results concerning the <u>asymptotic behavior of solutions</u>, which can be applied to equation (9.13). As special cases, we can obtain some results due to LaSalle [38], [41] and Levin and Nohel [43], [44].

<u>Definition 10.2.</u> A scalar function $W(x)$ defined for $x \varepsilon Q$ is said to be <u>positive definite with respect to a set S</u>, if $W(x) = 0$ for $x \varepsilon S$ and if corresponding to each $\varepsilon > 0$ and each compact set Q^* in Q, there exists a positive number $\delta(\varepsilon,Q^*)$ such that

$$W(x) \geq \delta(\varepsilon,Q^*) \quad \text{for} \quad x \varepsilon Q^* - N(\varepsilon,S).$$

Theorem 10.3. Suppose that there exists a nonnegative Liapunov function $V(t,x)$ on $I \times Q$ such that $\dot{V}_{(10.4)}(t,x) \leq -W(x)$, where $W(x)$ is positive definite with respect to a closed set Ω in the space Q. Moreover, suppose that $f(t,x)$ satisfies conditions (a) and (b) with respect to Ω. Then every bounded solution of (10.4) approaches the largest semi-invariant set of the system (10.6) contained in Ω as $t \to \infty$.

Proof. Let $x(t)$ be a bounded solution defined for $t \geq t_0$. It is sufficient to prove that $x(t)$ approaches Ω as $t \to \infty$. Then, by Theorem 10.1, we can see that $x(t)$ approaches the largest semi-invariant set of (10.6) as $t \to \infty$.

Since $x(t)$ is bounded, there exists a compact set $Q^* \subset Q$ such that $x(t) \in Q^*$ for all $t \geq t_0$. First of all, we shall show that

(c) given $\varepsilon > 0$, there exists a sequence $\{t_k\}$ such that $t_k \to \infty$ as $k \to \infty$ and $x(t_k) \in N(\varepsilon,\Omega)$. Suppose that there does not exist such a sequence. Then there is a $T \geq t_0$ such that $x(t) \bar{\in} N(\varepsilon,\Omega)$ for all $t \geq T$. Since $x(t) \in Q^*$, there exists a $\delta(\varepsilon,Q^*) > 0$ such that

$$W(x(t)) \geq \delta(\varepsilon,Q^*) \quad \text{for} \quad t \geq T,$$

which implies that

$$V(t,x(t)) - V(T,x(T)) \leq -\delta(\varepsilon,Q^*)(t-T) \to -\infty \quad (t \to \infty).$$

This contradicts $V(t,x(t)) \geq 0$. Thus we have (c).

Now we shall show that $\Omega \cap Q^* = Q_1$ is nonempty. For $\varepsilon = \frac{1}{m}$, it follows from (c) that there exists a t_m such that $x(t_m) \in N(\frac{1}{m},\Omega)$ and we can assume that $t_m \to \infty$ as $m \to \infty$. Since $x(t_m) \in Q^*$, there exists an $x_0 \in Q^*$ and a subsequence $\{t_{m_j}\}$ of $\{t_m\}$ such that

$x(t_{m_j}) \to x_0$ as $j \to \infty$. On the other hand,

$$\text{dist}(x_0,\Omega) \leq \text{dist}(x_0,x(t_{m_j})) + \text{dist}(x(t_{m_j}),\Omega)$$

and

$$x(t_{m_j}) \in N(1/m_j,\Omega),$$

and hence, we have $\text{dist}(x_0,\Omega) = 0$, that is, $x_0 \in \Omega$, because Ω is
closed. This shows that $Q_1 \neq \emptyset$.

By condition (b), for $y \in Q_1$ there exists a $\delta = \delta(Q_1)$ such
that if $|x-y| \leq \delta$ and $t \in I$, then $|f(t,x)-f(t,y)| \leq 1$. Moreover,
there exists a $T = T(Q_1)$ such that if $y \in Q_1$ and $t \geq T$, then
$|f(t,y)-h(y)| < 1$. Therefore, if $y \in Q_1$, $|x-y| \leq \delta$ and $t \geq T$,
we have $|f(t,x)| < |h(y)| + 2$. Thus we can find a constant
$K = K(Q_1) > 0$ such that $x \in N(\delta,Q_1)$ implies $|f(t,x)| \leq K$ for
large t.

Now suppose that $x(t)$ does not approach Ω. Then, for some
$\varepsilon > 0$ there exists a sequence $\{t_k\}$ such that $t_k \to \infty$ as $k \to \infty$
and $\text{dist}(x(t_k),\Omega) \geq \varepsilon$, where we can assume that $\varepsilon < \delta(Q_1)$. On the
other hand, by (c), there exists a sequence $\{\bar{t}_k\}$ such that $\bar{t}_k \to \infty$
as $k \to \infty$ and $d(x(\bar{t}_k),\Omega) < \frac{\varepsilon}{2}$. Thus we can find sequences $\{\tau_m\}$
and $\{\bar{\tau}_m\}$ such that

$$\tau_m < \bar{\tau}_m, \ \tau_m \to \infty \quad \text{as} \quad m \to \infty$$

and

$$\text{dist}(x(\tau_m),\Omega) = \varepsilon, \ \text{dist}(x(\bar{\tau}_m),\Omega) = \frac{\varepsilon}{2},$$

where we can assume that τ_1 is sufficiently large and the intervals
$\tau_m \leq t \leq \bar{\tau}_m$ are disjoint and

$$\frac{\varepsilon}{2} < \text{dist}(x(t),\Omega) < \varepsilon \quad \text{for} \quad \tau_m < t < \bar{\tau}_m.$$

Therefore there exists a $\gamma > 0$ such that $V'(t,x(t)) \leq -\gamma$ for
$t \in [\tau_m,\bar{\tau}_m]$. For large t and $x \in N(\delta,Q_1)$, $|f(t,x)| \leq K$, and if m

is sufficiently large,

$$\int_{\tau_m}^{\tau_m + \varepsilon/4K} |g(t,x(t))| dt < \frac{\varepsilon}{4}.$$

Therefore, if $\tau_m + \frac{\varepsilon}{4k} > \bar{\tau}_m$, we have $|x(t) - x(\tau_m)| < \frac{\varepsilon}{2}$ on the interval $\tau_m \le t \le \bar{\tau}_m$. From

$$\text{dist}(x(t), \Omega) \ge \text{dist}(x(\tau_m), \Omega) - |x(t) - x(\tau_m)|,$$

it follows that $\text{dist}(x(t), \Omega) > \frac{\varepsilon}{2}$. Therefore $\tau_m + \frac{\varepsilon}{4K} \le \bar{\tau}_m$, because $\text{dist}(x(\bar{\tau}_m), \Omega) = \frac{\varepsilon}{2}$. Thus

$$V(\bar{\tau}_m, x(\bar{\tau}_m)) - V(t_0, x(t_0)) \le -\gamma \frac{\varepsilon}{4K} m \to -\infty \quad \text{as} \quad m \to \infty,$$

which contradicts $V(t,x) \ge 0$. This shows that $x(t) \to \Omega$ as $t \to \infty$. Thus the proof is completed.

Example 10.1 [44]. In the equation

$$x'' + h(t,x,x')x' + f(x) = e(t), \qquad (10.16)$$

we assume that

   (i) $h(t,x,y)$ is continuous, nonnegative on $I \times R^1 \times R^1$ and is bounded when $x^2 + y^2$ is bounded, and moreover $h(t,x,y) \ge k(x,y) > 0$ for $y \ne 0$, where $k(x,y)$ is a continuous function,

   (ii) $f(x)$ is continuous in R^1, $xf(x) > 0$ for $x \ne 0$ and

$$F(x) = \int_0^x f(u)du \to \infty \quad \text{as} \quad |x| \to \infty,$$

   (iii) $e(t)$ is continuous on I and $E(t) = \int_0^t |e(s)| ds < \infty$.

Then every solution $x(t)$ of (10.16) exists in the future and $x(t) \to 0$, $x'(t) \to 0$ as $t \to \infty$.

To see this, consider a system equivalent to (10.16)

$$x' = y, y' = -h(t,x,y)y - f(x) + e(t) \qquad (10.17)$$

and a Liapunov function

$$V(t,x,y) = e^{-2E(t)} \{F(s) + \frac{y^2}{2} + 1\}.$$

Then we have

$$e^{-2E(\infty)} \{F(x) + \frac{y^2}{2} + 1\} \le V(t,x,y) \le F(x) + \frac{y^2}{2} + 1$$

and

$$\dot{V}_{(10.17)}(t,x,y) \le -e^{-2E(\infty)} h(t,x,y)y^2,$$

where $E(\infty) = \int_0^\infty |e(s)| ds$.

Then, by Theorem 8.8, the solutions of (10.17) are uniformly bounded. By the condition on $h(t,x,y)$ and (10.18), the set Ω in Theorem 10.3 is the set of all points where $y = 0$, that is, the x-axis. The other conditions in Theorem 10.3 can be easily verified. On the set Ω, the system (10.6) corresponds to

$$x' = 0, \quad y' = -f(x).$$

By the condition on $f(x)$, the largest semi-invariant set contained in Ω is only the origin. Therefore, by Theorem 10.3, $x(t) \to 0$, $y(t) \to 0$ as $t \to \infty$.

For some more discussion about this type equation, see [56], [67], [69].

LaSalle has extended his result to a periodic system [39] and Miller [49] has extended some of our results to a system

$$x' = p(t,x) + f(t,x) + g(t,x),$$

where $p(t,x)$ is almost periodic in t, $f(t,x) \to 0$ uniformly on any compact set in R^n as $t \to \infty$ and $g(t,x)$ is integrable in the sense of (10.5). For results in functional differential equations, see [27],

[29], [50]. For a discussion on this line, see [7], [65], [68], [72].

Now we shall consider a system

$$x' = f(t,x),   \qquad (10.19)$$

where $f(t,x)$ is continuous on $I \times Q$, Q: open set in R^n.

Theorem 10.4. We assume that there exists a nonnegative Liapunov function $V(t,x)$ defined on $I \times Q$ such that $\dot{V}_{(10.19)}(t,x) \le -W(x) \le 0$, where $W(x)$ is continuous on Q. Let E be the set of x such that $W(x) = 0$, and let $x(t)$ be a bounded solution of (10.19) for $t \ge t_0$, that is, there is a compact set Q^* in Q for which $x(t) \in Q^*$ for all $t \ge t_0$. Then, if

$$\varlimsup_{h \to 0^+} \frac{1}{h}\{W(x(t+h)) - W(x(t))\}   \qquad (10.20)$$

is bounded from above (or from below), then $x(t) \to E$ as $t \to \infty$.

Proof. Since $x(t) \in Q^*$ and $\dot{V}_{(10.19)}(t,x) \le -W(x)$, $V'(t,x(t)) \le -W(x(t))$ for all $t \ge t_0$. Thus we have

$$V(t,x(t)) - V(t_0,x(t_0)) \le -\int_{t_0}^{t} W(x(t))dt,$$

and hence

$$\int_{t_0}^{\infty} W(x(t))dt < \infty.   \qquad (10.21)$$

Suppose that $W(x(t))$ does not tend to zero as $t \to \infty$. Then, for some $\varepsilon > 0$, there exists a sequence $\{t_k\}$ such that $t_k \to \infty$ as $k \to \infty$ and that $W(x(t_k)) \ge \varepsilon$. In the case where (10.20) is bounded by $K > 0$ from above, $W(x(t)) \ge \frac{\varepsilon}{2}$ on the intervals

$$t_k - \frac{\varepsilon}{2K} \le t \le t_k.$$

In the case where (10.20) is bounded by $-L, L > 0$, from below, $W(x(t)) \ge \frac{\varepsilon}{2}$ on the intervals

$$t_k \le t \le t_k + \varepsilon/2L.$$

We can assume that these intervals are disjoint. Thus we have $\int_{t_0}^{\infty} W(x(t))dt = \infty$, which contradicts (10.21). Therefore $W(x(t)) \to 0$ as $t \to \infty$.

Let p be a positive limit point of x(t). Then there exists a sequence $\{\tau_k\}$ such that $\tau_k \to \infty$ and $x(\tau_k) \to p$ as $k \to \infty$, and hence $W(p) = 0$ since $W(x(\tau_k)) \to 0$ as $k \to \infty$ and $W(x)$ is continuous. Therefore p ε E and consequently $x(t) \to E$ as $t \to \infty$.

From the proof of Theorem 10.4, it is clear that we can state the theorem in the following way. Let x(t) be a solution of (10.19) such that $x(t) \varepsilon Q^*$ for all $t \ge t_0$, where Q^* is a compact set in Q. We assume that there exists a Liapunov function $V(t,x) \ge 0$ defined on $I \times Q$ such that $\dot{V}_{(10.19)}(t,x) \le -W(x) \le 0$ for $t \ge t_0$ and $x \varepsilon Q^*$, where $W(x)$ is continuous on Q^*. Let E be the set of x such that $W(x) = 0$, $x \varepsilon Q^*$. Then, if (10.20) is bounded from above (or from below), then $x(t) \to E$ as $t \to \infty$.

Lemma 10.6. For a given continuous function $W(x) \ge 0$ defined on Q^*, let E be the set of x such that $W(x) = 0$. Then there exists a continuous function $W^*(x) \ge 0$ defined on Q^* which has the following properties;

    (i) $W(x) \ge W^*(x)$,

    (ii) $|W^*(x) - W^*(y)| \le L|x-y|$, where $L > 0$ is a constant and $x \varepsilon Q^*$, $y \varepsilon Q^*$,

    (iii) the set of x such that $W^*(x) = 0$ is also E.

Proof. Let $r = \text{dist}(x,E)$, $x \varepsilon Q^*$. Since $W(x)$ is continuous on the compact set Q^* and since $W(x) = 0$ for $x \varepsilon E$ and $W(x) > 0$ for $x \bar{\varepsilon} E$, there is a continuous increasing function $W_1(r) \ge 0$ such that

$$W_1(0) = 0, \quad W_1(r) > 0 \quad \text{for} \quad r > 0$$

and

$$W_1(r) \leq W(x) \quad \text{for} \quad x \, \varepsilon \, Q^*.$$

Defining $W_1(r) = 0$ for $r < 0$, set $W_2(r) = \int_{r-1}^{r} W_1(s)\,ds$. Then we have

$$W_2(0) = 0, \quad W_2(r) > 0 \quad \text{for} \quad r > 0, \quad W_2(r) \leq W_1(r) \leq W(x)$$
$$\text{for} \quad x \, \varepsilon \, Q^*$$

and

$$W_2'(r) = W_1(r) - W_1(r-1) \geq 0,$$

and hence, there is an $L > 0$ such that

$$|W_2(r) - W_2(\bar{r})| \leq L|r - \bar{r}|.$$

Let $W^*(x) = W_2(\text{dist}(x,E))$ for $x \, \varepsilon \, Q^*$. Then $W^*(x) = 0$ for $x \, \varepsilon \, E$, $W^*(x) > 0$ for $x \, \bar{\varepsilon} \, E$, $W^*(x) \leq W(x)$ on Q^* and $|W^*(x) - W^*(y)| \leq L|x-y|$.

If there exists a Liapunov function $V(t,x) \geq 0$ on $I \times Q$ and $\dot{V}_{(10.19)}(t,x) \leq -W(x) \leq 0$ for $x \, \varepsilon \, Q^*$ and $t \geq t_0$, then clearly $\dot{V}_{(10.19)}(t,x) \leq -W^*(x) \leq 0$ for $t \geq t_0$ and $x \, \varepsilon \, Q^*$. Thus, if $f(t,x)$ is bounded for $t \geq t_0$ and $x \, \varepsilon \, Q^*$, that is, $|f(t,x)| \leq K$, $t \geq t_0$, $x \, \varepsilon \, Q^*$ for some constant K,

$$\overline{\lim_{h \to 0^+}} \frac{1}{h}\{W^*(x(t+h)) - W^*(x(t))\}$$

$$\leq \overline{\lim_{h \to 0^+}} \frac{1}{h} L|x(t+h) - x(t)|$$

$$\leq L|f(t,x)| \leq LK,$$

and hence, we can apply the above result and we can conclude that $x(t) \to E$ as $t \to \infty$.

The following is a more general discussion for a nonautonomous

system.   Following LaSalle's paper [40], consider a nonautonomous
system

$$x' = f(t,x). \tag{10.22}$$

Let G be a set in R^n and let Q be an open set of R^n containing
the closure $\overline{G}$ of G. We assume that $f(t,x) \in C(I \times Q, R^n)$. Let
$V(t,x)$ be a Liapunov function defined on $I \times Q$. Now let $x(t)$ be
a solution of (10.22) that remains in G for $t \geq 0$ and let $[0,\omega)$
be its maximal positive interval of definition (ω can be ∞).

In what follows, we say that $V(t,x)$ <u>is a Liapunov function</u>
<u>for</u> (10.22) <u>on</u> G, if it is continuous and locally Lipschitzian in x
on $I \times Q$ and if

   (i)   given $x \in \overline{G}$, there is a neighborhood N of x such
         that $V(t,x)$ is bounded from below for all $t \geq 0$ and
         all $x \in N \cap G$,

   (ii) $\dot{V}_{(10.22)}(t,x) \leq -W(x) \leq 0$ for all $t \geq 0$ and all $x \in G$,
         where $W(x)$ is continuous on $\overline{G}$.

We now compactify the space R^n and denote the one-point
compactification of R^n by R^n_∞. Let $d(x,y) = |x-y|$ denote the
Euclidian distance between x and y, and define $d(x,\infty) = \dfrac{1}{|x|}$. For
S a set in R^n_∞, define $d(x,S) = \inf\{d(x,y); y \in S\}$. Let $x(t)$ be
a continuous function on $[0,\omega)$. Then $x(t) \to S$ as $t \to \omega^-$ means
$d(x(t),S) \to 0$ as $t \to \omega^-$. Let $V(t,x)$ be a Liapunov function for
(10.22) on G and define

$$E = \{x; W(x) = 0, x \in \overline{G}\} \quad \text{and} \quad E_\infty = E \cup \{\infty\}.$$

<u>Theorem 10.5.</u> Let $V(t,x)$ be a Liapunov function for (10.22)
on G, and let $x(t)$ be a solution of (10.22) which remains in G

for $t \geq t_0 \geq 0$ with $[t_0, \omega)$ the maximal future interval of definition of $x(t)$.

(a)   If for each $p \in \bar{G}$ there is a neighborhood N of p
such that $|f(t,x)|$ is bounded for all $t \geq 0$ and all
$x \in N \cap G$, then either $x(t) \to \infty$ as $t \to \omega^-$, or $\omega = \infty$
and $x(t) \to E_\infty$ as $t \to \infty$.

(b)   If

$$\overline{\lim_{h \to 0^+}} \frac{1}{h} \{W(x(t+h)) - W(x(t))\} \qquad (10.23)$$

is bounded from above (or from below) on $[t_0, \omega)$ and if
$\omega = \infty$, then $x(t) \to E_\infty$ as $t \to \infty$.

Proof. Let $p \in R^n$ be a finite positive limit point of $x(t)$.
Then there is an increasing sequence $\{t_k\}$ such that $t_k \to \omega^-$ and
$x(t_k) \to p$ as $k \to \infty$. Since $p \in \bar{G}$, there is a neighborhood N of p
such that $V(t,x)$ is bounded from below for all $t \geq 0$ and all
$x \in N \cap G$. Since $x(t_k) \in N \cap G$ for sufficiently large k ,
$V(t_k, x(t_k))$ is bounded from below. On the other hand,
$\dot{V}_{(10.22)}(t,x) \leq 0$ for all $t \geq 0$ and all $x \in G$. Therefore $V(t,x($
$V(t,x(t))$ is nonincreasing, and hence $V(t_k, x(t_k)) \to c$ as $k \to \infty$
for some constant c and consequently $V(t,x(t)) \to c$ as $t \to \omega^-$.
By condition (ii), we have

$$V(t,x(t)) - V(t_0, x(t_0)) \leq -\int_{t_0}^{t} W(x(s)) ds$$

on $t_0 \leq t < \omega$, and hence

$$\int_{t_0}^{\omega} W(x(s)) ds < \infty. \qquad (10.24)$$

We now prove (a). Let p be a finite positive limit point of
$x(t)$ and assume that p is not in E . Then there is a $\delta > 0$ such
that $W(p) > 2\delta > 0$. Since $W(x)$ is continuous on $\bar{G}$, there exists

a neighborhood $N(2\varepsilon,p)$ of radius 2ε about p such that $W(x) > \delta$
for $x \in N(2\varepsilon,p) \cap \bar{G}$. If $x(t)$ remains in $N(2\varepsilon,p)$ for all
$t \in [t_1,\omega)$ for some $t_1 \geq t_0$, ω would be ∞ and

$$\int_{t_0}^{t_1} W(x(s))ds + \int_{t_1}^{\infty} W(x(s))ds \geq \int_{t_0}^{t_1} W(x(s))ds + \int_{t_1}^{\infty} \delta ds,$$

which contradicts (10.24).

The other possibility is that $x(t)$ goes in and out of
$N(2\varepsilon,p)$ an infinite number of times. This means that $x(t)$ travels
an infinite distance within $N(2\varepsilon,p)$ since p is a positive limit
point and $x(t)$ must enter $N(\varepsilon,p)$ an infinite number of times. For
sufficiently small ε, $|x'(t)|$ is bounded in $N(2\varepsilon,p)$, and hence
$x(t)$ must remain in $N(2\varepsilon,p)$ an infinite length of time. This implies $\omega = \infty$, and this contradicts (10.24). Therefore $p \in E$ and
$W(p) = 0$. Namely E contains all finite positive limit points. The
above also shows that if $x(t)$ has finite positive limit points, then
$\omega = \infty$, and this completes the proof of (a).

Next we shall prove (b). We here assume $\omega = \infty$. Let p be
a finite positive limit point of $x(t)$. Then there is a sequence
$\{t_k\}$ such that $t_k \to \infty$ as $k \to \infty$ and $x(t_k) \to p$ as $k \to \infty$. We
show that $W(x(t_k)) \to 0$ as $k \to \infty$. Suppose not. Then there exists
an $\varepsilon > 0$ and a subsequence of $\{t_k\}$, which we shall denote by $\{t_k\}$
again, such that $t_k \to \infty$ as $k \to \infty$ and $W(x(t_k)) \geq \varepsilon$. In the case
where (10.23) is bounded by $K > 0$ from above, we have

$$W(x(t)) \geq \frac{\varepsilon}{2} \quad \text{on} \quad t_k - \frac{\varepsilon}{2K} \leq t \leq t_k.$$

Therefore $\displaystyle\int_{t_0}^{\infty} W(x(s))ds \geq \frac{\varepsilon}{2} \cdot \frac{\varepsilon}{2K} \cdot k$, where we can assume that
$t_1 - \frac{\varepsilon}{2K} \geq t_0$ and these intervals are disjoint. Letting $k \to \infty$,
$\displaystyle\int_{t_0}^{\infty} W(x(s))ds = \infty$, which contradicts (10.24). In the case where (10.23)

is bounded by $-L, L > 0$, from below, consider intervals $t_k \leq t \leq$ $t_k + \frac{\varepsilon}{2L}$. Thus we see that $W(x(t_k)) \to 0$ as $k \to \infty$. Since $W(x)$ is continuous on $\overline{G}$, $W(p) = 0$, that is, $p \in E$. This proves that $x(t) \to E_\infty$ as $t \to \infty$.

Remark 1. LaSalle has stated Theorem 10.5(b) in the following way: If $W(x(t))$ is absolutely continuous and its derivative is bounded from above (or from below) almost everywhere on $[t_0, \omega)$ and if $\omega = \infty$, then $x(t) \to E_\infty$ as $t \to \infty$. But if $W(x(t)) = f(t)$ is absolutely continuous and its derivative is bounded by K from above almost everywhere, we have

$$f(t+h) - f(t) = \int_t^{t+h} f'(s)ds \leq Kh, \quad h > 0,$$

and hence (10.23) is bounded by K from above.

Remark 2. If V_1 and V_2 are Liapunov functions for (10.22) on G, then $V = V_1 + V_2$ is also a Liapunov function for (10.22) on G and $E_\infty = E_\infty^1 \cap E_\infty^2$.

Remark 3. If E is made up of a number of components (maximal connected sets) and $x(t)$ remains in G for $t \geq 0$, then $x(t)$ approaches just one of these components since the positive limit set of $x(t)$ is connected. For example, if E is bounded, then either $x(t) \to \infty$ as $t \to \omega^-$ or $x(t) \to E$ as $t \to \infty$.

The proof of the following theorem is contained in the proof of Theorem 10.5.

Theorem 10.6. If the Liapunov function in Theorem 10.5 does not depend on t, then "$x(t) \to E_\infty$ as $t \to \infty$" can be replaced by "$x(t) \to (E \cap Q_c) \cup \{\infty\}$" for some c, where $Q_c = \{x; V(x) = c\}$.

Example 10.2. Consider a system

$$x' = y, \quad y' = -p(t)y - x, \qquad\qquad (10.25)$$

where $p(t)$ is continuous and $p(t) \geq \delta > 0$.

Using a Liapunov function $V(x,y) = x^2 + y^2$, we can see that the solutions of (10.25) are uniformly bounded, because

$$\dot{V}_{(10.25)}(x,y) = 2xy - 2p(t)y^2 - 2xy \leq -2\delta y^2.$$

If $p(t)$ is bounded, it follows from Theorem 10.3 that every solution of (10.25) approaches zero as $t \to \infty$. But if $p(t)$ is not bounded, we can not apply Theorem 10.3 and also Theorem 10.5(a). However, in this example, $W = 2\delta y^2$ and hence

$$\begin{aligned}
W'(y(t)) &= 4\delta y(t)[-p(t)y(t) - x(t)] \\
&= -4\delta p(t)y^2(t) - 4\delta x(t)y(t) \\
&\leq -4\delta x(t)y(t).
\end{aligned}$$

As was seen, the solutions of (10.25) are uniformly bounded, and hence $W'(y(t))$ is bounded from above. Thus, by Theorem 10.5(b), we can see that $y(t) \to 0$ as $t \to \infty$. Moreover, since V does not depend on t, we can conclude that $x(t) \to$ constant, $y(t) \to 0$ as $t \to \infty$.

Noting that the equation

$$x'' + (2+e^t)x' + x = 0$$

has a solution $x(t) = 1 + e^{-t}$, we see that this is the best possible result without further restrictions on $p(t)$.

11.   Converse Theorems

For the linear system

$$x' = A(t)x, \qquad\qquad (11.1)$$

where $A(t)$ is an $n \times n$ continuous matrix on I, we have the following converse theorem.

Theorem 11.1. Suppose that there exists a $K > 0$ and a constant c such that

$$|x(t,t_0,x_0)| \le Ke^{-c(t-t_0)}|x_0|, \qquad (11.2)$$

where $x(t,t_0,x_0)$ is a solution of (11.1). Then there exists a Liapunov function $V(t,x)$ which satisfies the following conditions;

(i) $|x| \le V(t,x) \le K|x|$,

(ii) $|V(t,x) - V(t,y)| \le K|x-y|$,

(iii) $\dot{V}_{(11.1)}(t,x) \le -cV(t,x)$.

This Liapunov function $V(t,x)$ is defined by

$$V(t,x) = \sup_{\tau \ge 0}|x(t+\tau,t,x)|e^{c\tau}.$$

For the proof, see [28], [80].

Remark. In this theorem, c is not necessarily positive. If $c > 0$, the zero solution is uniformly asymptotically stable. If $c = 0$, the zero solution is uniformly stable.

Now we shall discuss converse theorems on asymptotic stability in the large of the system

$$x' = f(t,x), \qquad (11.3)$$

where $f(t,x) \in C(I \times R^n, R^n)$ and $f(t,0) \equiv 0$. First of all, we state a simple form of a lemma due to Massera [47]. For the proof, see [80].

Lemma 11.1. Given any real function $A(r,t,\varepsilon)$ of real variables, defined, continuous and positive in $Q: r \ge 0$, $t \ge 0$, $\varepsilon > 0$, there exist three continuous functions $h(r), p(t), g(\varepsilon)$ such that $h(r) > 0$, $p(t) > 0$, $g(\varepsilon) > 0$ for $\varepsilon > 0$, $g(0) = 0$ and that

$$h(r)p(t)g(\varepsilon) \leq A(r,t,\varepsilon) \quad \text{in} \quad Q. \tag{11.4}$$

Theorem 11.2. We assume that $f(t,x)$ is locally Lipschitzian in x. If the zero solution of (11.3) is equiasymptotically stable in the large, there exists a Liapunov function $V(t,x)$ defined on $I \times R^n$ which satisfies the following conditions;

(i) $V(t,0) \equiv 0$,

(ii) $a(|x|) \leq V(t,x)$, where $a(r)$ is continuous, increasing, positive definite and $a(r) \to \infty$ as $r \to \infty$,

(iii) $|V(t,x) - V(t,y)| \leq h(\alpha)p(t)|x-y|$ for $|x| \leq \alpha$ and $|y| \leq \alpha$, where $h(\alpha),p(t)$ are suitable continuous functions,

(iv) $\overset{\cdot}{V}_{(11.3)}(t,x) \leq -cV(t,x)$, where $c > 0$ is a constant.

Proof. Let $\Omega_{\sigma,\alpha}$ be a domain such that $0 \leq t \leq \sigma$, $|x| \leq \alpha$. Since equiasymptotic stability in the large implies the equi-boundedness of solutions of (11.3), there exists a $\beta(\sigma,\alpha) > 0$ such that if $(t_0,x_0) \in \Omega_{\sigma,\alpha}$, then

$$|x(t,t_0,x_0)| \leq \beta(\sigma,\alpha) \quad \text{for all} \quad t \geq t_0.$$

Moreover, there is a $T(\sigma,\alpha,\varepsilon) > 0$ such that $(t_0,x_0) \in \Omega_{\sigma,\alpha}$ implies that

$$|x(t,t_0,x_0)| < \varepsilon \quad \text{for all} \quad t \geq t_0 + T(\sigma,\alpha,\varepsilon).$$

If $\varepsilon > 1$, we set $T(\sigma,\alpha,\varepsilon) = T(\sigma,\alpha,1)$. Since $f(t,x)$ is locally Lipschitzian in x, there exists an $L(\sigma,\alpha,\varepsilon) > 0$ such that if $0 \leq t \leq \sigma + T(\sigma,\alpha,\varepsilon)$, $|x| \leq \beta(\sigma,\alpha)$ and $|y| \leq \beta(\sigma,\alpha)$, then

$$|f(t,x)-f(t,y)| \leq L(\sigma,\alpha,\varepsilon)|x-y|.$$

Let $F(\sigma,\alpha,\varepsilon)$ be $1 + \max|f(t,x)|$ for $0 \leq t \leq \sigma + T(\sigma,\alpha,\varepsilon)$,

$|x| \leq \beta(\sigma,\alpha)$.   Furthermore, β, T, L and F can be assumed to be continuous.   For a given $c > 0$, let

$$A(\sigma,\alpha,\epsilon) = e^{cT(\sigma,\alpha,\epsilon)} \{2F(\sigma,\alpha,\epsilon)e^{L(\sigma,\alpha,\epsilon)T(\sigma,\alpha,\epsilon)} + \beta(\sigma,\alpha)\}. \quad (11.5)$$

Then, by Lemma 11.1, there exist continuous functions $h(\alpha), p(\sigma)$ and $g(\epsilon)$ such that $h(\alpha) > 0$, $p(\sigma) > 0$, $g(\epsilon) > 0$ for $\epsilon > 0$, $g(0) = 0$ and that

$$g(\epsilon)A(\sigma,\alpha,\epsilon) \leq p(\sigma)h(\alpha). \quad (11.6)$$

For $k = 1,2,\ldots$, we define $V_k(t,x)$ by

$$V_k(t,x) = g(\tfrac{1}{k}) \sup_{\tau \geq 0} G_k(|x(t+\tau,t,x)|)e^{c\tau}, \quad (11.7)$$

where

$$G_k(z) = \begin{cases} z - \dfrac{1}{k} & (z \geq \dfrac{1}{k}) \\[2mm] 0 & (0 \leq z < \dfrac{1}{k}). \end{cases} \quad (11.8)$$

Clearly $G_k(z) \to \infty$ as $z \to \infty$ for each k and

$$|G_k(z) - G_k(z')| \leq |z - z'|.$$

From the definition of $V_k(t,x)$, it is clear that

$$V_k(t,0) \equiv 0, \quad (11.9)$$

$$g(\tfrac{1}{k})G_k(|x|) \leq V_k(t,x) \quad (11.10)$$

and if $(t,x) \in \Omega_{\sigma,\alpha}$,

$$V_k(t,x) \leq g(\tfrac{1}{k})G_k(\beta(\sigma,\alpha))e^{cT(\sigma,\alpha,1/k)} \quad (11.11)$$

$$\leq g(\tfrac{1}{k})\beta(\sigma,\alpha)e^{cT(\sigma,\alpha,1/k)}.$$

Therefore, by (11.5) and (11.6),

$$V_k(t,x) \leq p(\sigma)h(\alpha) \quad \text{for} \quad (t,x) \in \Omega_{\sigma,\alpha}. \quad (11.12)$$

Now we shall show that $V_k(t,x)$ is locally Lipschitzian in t and x. For $(t,x) \in \Omega_{\sigma,\alpha}$ and $(s,y) \in \Omega_{\sigma,\alpha}$, $t < s$,

$$|V_k(t,x)-V_k(s,y)| \leq g(\tfrac{1}{k}) \sup_{\tau \geq 0} |G_k(|x(t+\tau,t,x)|)-G_k(|x(s+\tau,s,y)|)| e^{c\tau}$$

$$\leq g(\tfrac{1}{k}) \sup_{T(\sigma,\alpha,\frac{1}{k}) \geq \tau \geq 0} e^{c\tau} |x(t+\tau,t,x)-x(s+\tau,s,y)|.$$

Letting $\xi = x(s,t,x)$, we have

$$|x(t+\tau,t,x) - x(s+\tau,s,y)|$$

$$\leq |x(t+\tau,t,x)-x(s+\tau,t,x)| + |x(s+\tau,t,x)-x(s+\tau,s,y)|$$

$$\leq F(\sigma,\alpha,\tfrac{1}{k})(s-t) + |x(s+\tau,s,\xi)-x(s+\tau,s,y)|.$$

On the other hand, for $0 \leq \tau \leq T(\sigma,\alpha,\tfrac{1}{k})$

$$|x(s+\tau,s,\xi)-x(s+\tau,s,y)| \leq |\xi-y| e^{L(\sigma,\alpha,\frac{1}{k})T(\sigma,\alpha,\frac{1}{k})}$$

$$\leq \{|\xi-x|+|x-y|\} e^{L(\sigma,\alpha,\frac{1}{k})T(\sigma,\alpha,\frac{1}{k})}$$

$$\leq \{F(\sigma,\alpha,\tfrac{1}{k})(s-t)+|x-y|\} e^{L(\sigma,\alpha,\frac{1}{k})T(\sigma,\alpha,\frac{1}{k})}.$$

Thus we have

$$|V_k(t,x)-V_k(s,y)|$$

$$\leq g(\tfrac{1}{k}) 2F(\sigma,\alpha,\tfrac{1}{k}) e^{(c+L(\sigma,\alpha,\frac{1}{k}))T(\sigma,\alpha,\frac{1}{k})} \{(s-t)+|x-y|\},$$

and hence, by (11.5) and (11.6),

$$|V_k(t,x)-V_k(s,y)| \leq p(\sigma)h(\alpha)\{|s-t|+|x-y|\}. \tag{11.13}$$

Next we shall prove that $\dot{V}_{k(11.3)}(t,x) \leq -cV_k(t,x)$. For $\delta > 0$ and $\xi = x(t+\delta,t,x)$,

$$V_k(t+\delta,\xi) = g(\tfrac{1}{k}) \sup_{\tau \geq 0} G_k(|x(t+\delta+\tau,t+\delta,\xi)|) e^{c\tau}$$

$$= g(\tfrac{1}{k}) \sup_{\tau \geq 0} G_k(|x(t+\delta+\tau,t,x)|) e^{c\tau}$$

$$= g(\frac{1}{k}) \sup_{\tau \geq \delta} G_k(|x(t+\tau,t,x)|) e^{c\tau} e^{-c\delta}$$

$$\leq V_k(t,x) e^{-c\delta},$$

which implies

$$\dot{V}_{k(11.3)}(t,x) \leq -c V_k(t,x). \tag{11.14}$$

We now define $V(t,x)$ by

$$V(t,x) = \sum_{k=1}^{\infty} \frac{1}{2^k} V_k(t,x). \tag{11.15}$$

Since (11.12) implies the uniform convergence of the series in (11.15) on $\Omega_{\sigma,\alpha}$ and σ,α are arbitrary, $V(t,x)$ is defined and continuous on $I \times R^n$. Clearly $V(t,0) \equiv 0$. For x such that $|x| \geq 1$, by (11.10) and (11.15),

$$V(t,x) > \frac{1}{2} V_1(t,x) \geq \frac{1}{2} g(1) G_1(|x|) \geq \frac{1}{2} g(1)(|x|-1)$$

and for x such that $\frac{1}{k} \leq |x| < \frac{1}{k-1}$,

$$V(t,x) \geq \frac{1}{2^{k+1}} V_{k+1}(t,x) \geq \frac{1}{2^{k+1}} g(\frac{1}{k+1}) G_{k+1}(|x|)$$

$$\geq \frac{1}{2^{k+1}} g(\frac{1}{k+1})(|x| - \frac{1}{k+1})$$

$$\geq \frac{1}{2^{k+1}} g(\frac{1}{k+1}) \frac{1}{k(k+1)}.$$

Therefore we can find an $a(r)$ which is continuous, increasing, positive definite and satisfies the conditions that $a(r) \to \infty$ as $r \to \infty$ and $a(|x|) \leq V(t,x)$.

For $(t,x) \in \Omega_{\sigma,\alpha}$ and $(s,y) \in \Omega_{\sigma,\alpha}$,

$$|V(t,x)-V(s,y)| \leq \sum_{k=1}^{\infty} 1/2^k |V_k(t,x)-V_k(s,y)|$$

$$\leq \sum_{k=1}^{\infty} 1/2^k p(\sigma) h(\alpha) \{|t-s| + |x-y|\}$$

$$\leq p(\sigma) h(\alpha) \{|t-s| + |x-y|\},$$

which implies (iii).

Finally we shall prove condition (iv). In fact, we have

$$\dot{V}_{(11.3)}(t,x) = \overline{\lim_{h\to 0^+}} \frac{1}{h}\{\sum_{k=1}^{\infty} \frac{1}{2^k}V_k(t+h,x+hf(t,x)) - \sum_{k=1}^{\infty} \frac{1}{2^k} V_k(t,x)\}$$

$$\leq \sum_{k=1}^{\infty} \overline{\lim_{h\to 0^+}} \frac{1}{2^k} \frac{1}{h}\{V_k(t+h,x+hf(t,x)) - V_k(t,x)\}$$

$$\leq \sum_{k=1}^{\infty} \frac{1}{2^k}(-cV_k(t,x)) = -cV(t,x).$$

If the zero solution of (11.3) is uniformly asymptotically stable in the large, β and T in the proof of Theorem 11.2 can be replaced by $\beta(\alpha)$ and $T(\alpha,\varepsilon)$, and we can assume that $\beta(\alpha) \to 0$ as $\alpha \to 0$, because the zero solution is uniformly stable. From (11.11) and $e^{cT(\alpha,\varepsilon)} \leq A(0,\alpha,\varepsilon)$, it follows that

$$V_k(t,x) \leq g(\frac{1}{k})A(0,\alpha,\varepsilon)\beta(\alpha) \leq p(0)h(\alpha)\beta(\alpha),$$

and hence, $V(t,x) \leq p(0)h(\alpha)\beta(\alpha)$ for all $t \in I$ and $|x| \leq \alpha$. This implies the existence of a function $b(r)$ which is continuous, increasing and satisfies $b(r) \to 0$ as $r \to 0$ and $V(t,x) \leq b(|x|)$. Thus we have the following <u>converse theorem on uniformly asymptotic stability in the large</u>.

Theorem 11.3. If $f(t,x)$ is locally Lipschitzian in x and if the zero solution of (11.3) is uniformly asymptotically stable in the large, there exists a Liapunov function $V(t,x)$ defined on $I \times R^n$ which satisfies the following conditions;

(i) $a(|x|) \leq V(t,x) \leq b(|x|)$, where $a(r)$, $b(r)$ are continuous, increasing, positive definite and $a(r) \to \infty$ as $r \to \infty$,

(ii) $|V(t,x)-V(t,y)| \leq p(t)h(\alpha)|x-y|$ for $|x| \leq \alpha$ and $|y| \leq \alpha$, where $p(t)$ and $h(\alpha)$ are suitable continuous functions,

(iii) $\dot{V}_{(11.3)}(t,x) \leq -c(|x|)$, where $c(r)$ is continuous and
positive definite.

In particular, condition (iii) can be replaced by $\dot{V}_{(11.3)}(t,x) \leq$
$-cV(t,x)$.

Remark.   As is clear from the proof of Theorem 11.2, if for
any $\alpha > 0$

$$|f(t,x)-f(t,y)| \leq L(\alpha)|x-y| \quad \text{for all} \quad t \varepsilon I,$$
$$|x| \leq \alpha \quad \text{and} \quad |y| \leq \alpha, \tag{11.16}$$

then L and F also depend only on α, and hence, for a suitable
function $h(\alpha)$ we have

$$|V(t,x)-V(t,y)| \leq h(\alpha)|x-y| \quad \text{for} \quad |x| \leq \alpha, \; |y| \leq \alpha. \tag{11.17}$$

From the proof of Theorem 11.2, converse theorems on equiasymptotic and uniformly asymptotic stability follow immediately.
Let $f(t,x)$ in (11.3) be assumed to be continuous on $I \times S$,
$S = \{x; |x| < H\}$, and let H_1 be a positive number such that $H_1 < \delta_0$,
where $\delta_0 > 0$ is the one for uniformly asymptotic stability. Considering only α such that $\alpha = H_1$ in Theorem 11.3, the following
theorem can be obtained.

Theorem 11.4.   Suppose that $f(t,x)$ in (11.3) is continuous
on $I \times S$ and $f(t,0) \equiv 0$. If $f(t,x)$ is locally Lipschitzian in
x and if the zero solution of (11.3) is uniformly asymptotically
stable, there exists a Liapunov function $V(t,x)$ defined on
$0 \leq t < \infty, \; |x| < H_1$ which satisfies the conditions in Theorem 7.9 and

$$|V(t,x) - V(t,y)| \leq p(t)|x-y|, \tag{11.18}$$

where p(t) is a suitable continuous function. Moreover, if f(t,x)

satisfies (11.16), condition (11.18) is replaced by

$$|V(t,x)-V(t,y)| \leq K|x-y| \quad \text{for some constant } K > 0. \qquad (11.19)$$

Now let us consider the case where f(t,x) is almost periodic

in t. We shall assume that

  (a) f(t,x) in (11.3) is defined on R × S, f(t,0) ≡ 0 and

       f(t,x) is almost periodic in t uniformly for x ε S,

  (b) f(t,x) satisfies the Lipschitz condition (11.16).

Theorem 11.5. Under the assumptions above, if the zero solu-

tion of the almost periodic system (11.3) is uniformly asymptotically

stable for $t \geq 0$, then there exists a Liapunov function V(t,x) de-

fined on $-\infty < t < \infty$, $|x| < H_1$, where $H_1 < H$ is a suitable constant,

which is almost periodic in t uniformly for $x \in S_{H_1}$, $S_{H_1} =$

$\{x;\ |x| < H_1\}$, and which satisfies (11.19) for all $t \in R$ and the

following conditions;

  (i) $a(|x|) \leq V(t,x) \leq b(|x|)$, where a(r), b(r) are continu-

        ous, increasing, and positive definite.

  (ii) $\dot{V}_{(11.3)}(t,x) \leq -cV(t,x)$, where c > 0 is a constant.

In particular, if f(t,x) is periodic in t of period ω, so

is V(t,x). If f is autonomous, we can find an autonomous Liapunov

function.

In this case, the assumptions imply that for any ε > 0, there

exists a δ(ε) > 0 and a T(ε) > 0 such that $|x_0| < \delta(\varepsilon)$ and

$-\infty < t_0 < \infty$ imply $|x(t,t_0,x_0)| < \varepsilon$ for all $t \geq t_0$ and $|x_0| < \delta_0$

implies $|x(t,t_0,x_0)| < \varepsilon$ for all $t \geq t_0 + T(\varepsilon)$. Therefore we can

prove the theorem by the same idea as in the proof of Theorem 11.4.

For the details, see [80]. For uniformly asymptotic stability for all

$t_0 \varepsilon$ R, we have a further result.

Lemma 11.2. Consider an almost periodic system

$$x' = f(t,x), \qquad\qquad (11.20)$$

where $f(t,x) \varepsilon C(R \times D, R^n)$, $D = \{x; |x| < B^*\}$, and $f(t,x)$ is almost periodic in t uniformly for $x \varepsilon D$. Let $\phi(t)$ be a solution of (11.20) defined on $0 \le t < \infty$ such that $|\phi(t)| \le B < B^*$ for all $t \ge 0$. Furthermore, we assume that the solutions of (11.20) are unique for initial value problem.

Then, if $\phi(t)$ is uniformly stable, there exists a solution $\psi(t)$ of (11.20) defined on R which is uniformly stable for all $t_0 \varepsilon$ R, that is, for any $\varepsilon > 0$ and any $t_0 \varepsilon$ R, there exists a $\delta(\varepsilon) > 0$ such that $|x_0| < \delta(\varepsilon)$ implies $|x(t, t_0, x_0)| < \varepsilon$ for all $t \ge t_0$. Moreover, if $\phi(t)$ is uniformly asymptotically stable, $\psi(t)$ is uniformly asymptotically stable for all $t_0 \varepsilon$ R.

Proof. Let S be the set of x such that $|x| \le \dfrac{B^*+B}{2}$. Then S is a compact set in D. There exists a sequence $\{\tau_k\}$ such that $\tau_k \to \infty$ and $f(t+\tau_k, x)$ converges to $f(t,x)$ uniformly on $R \times S$ as $k \to \infty$. If we set $\phi_k(t) = \phi(t+\tau_k)$, $\phi_k(t)$ is a solution of

$$x' = f(t+\tau_k, x) \qquad\qquad (11.21)$$

through $(0, \phi(\tau_k))$. For any α, $-\infty < \alpha < 0$, if k is sufficiently large, $\phi_k(t)$ is defined on $\alpha \le t$ and $|\phi_k(t)| \le B$ for all $t \ge \alpha$. Moreover, since $\phi(t)$ is uniformly stable, $\phi_k(t)$ is a uniformly stable solution of (11.21) on $\alpha \le t < \infty$ with the same pair $(\varepsilon, \delta(\varepsilon))$ as the one for $\phi(t)$. Thus $\{\phi_k(t)\}$ is uniformly bounded and equicontinuous on $\alpha \le t < \infty$, and hence there exists a subsequence of $\{\phi_k(t)\}$, which we shall denote by $\{\phi_k(t)\}$ again, such that $\phi_k(t)$ converges to a solution $\psi(t)$ of (11.20) defined on $-\infty < t < \infty$

uniformly on any compact interval in R.

For a fixed $t_0 \in R$, if k is sufficiently large, we have

$$\left|\phi_k(t_0) - \psi(t_0)\right| < \frac{1}{2}\,\delta\left(\frac{\varepsilon}{2}\right),\qquad\qquad (11.22)$$

where $\delta(\varepsilon)$ is the one for the uniform stability for $\phi(t)$ and we
can assume that $\varepsilon < B^* - B$. Let y_0 be such that

$$\left|y_0 - \psi(t_0)\right| < \frac{1}{2}\,\delta\left(\frac{\varepsilon}{2}\right)\qquad\qquad (11.23)$$

and let x(t) be the solution of (11.20) such that $x(t_0+\tau_k) = y_0$.
Then $x_k(t) = x(t+\tau_k)$ is the solution of (11.21) and $x_k(t_0) = y_0$.
Since $\phi_k(t)$ is uniformly stable and $\left|\phi_k(t_0)-x_k(t_0)\right| < \delta\left(\frac{\varepsilon}{2}\right)$ by
(11.22) and (11.23), we have

$$\left|\phi_k(t) - x_k(t)\right| < \frac{\varepsilon}{2} \text{ for all } t \geq t_0.\qquad\qquad (11.24)$$

Since $\left|x_k(t)\right| < B + \frac{\varepsilon}{2}$ for all $t \geq t_0$, the sequence $\{x_k(t)\}$ con-
verges to the solution y(t) of (11.20) through (t_0,y_0), which is
uniquely determined, uniformly on any compact interval $[t_0,t_0+N]$.
Thus, if k is sufficiently large, say $k \geq k_0(\varepsilon,N)$,

$$\left|x_k(t) - y(t)\right| < \frac{\varepsilon}{4} \text{ and } \left|\phi_k(t) - \psi(t)\right| < \frac{\varepsilon}{4}$$

$$\text{on } t_0 \leq t \leq t_0+N.\qquad\qquad (11.25)$$

From (11.24) and (11.25), it follows that

$$\left|\psi(t) - y(t)\right| < \varepsilon \text{ on } t_0 \leq t \leq t_0+N.$$

Since N is arbitrary, $\left|\psi(t)-y(t,t_0,y_0)\right| < \varepsilon$ for all $t \geq t_0$ if
$\left|\psi(t_0)-y_0\right| < \frac{1}{2}\,\delta\left(\frac{\varepsilon}{2}\right)$, where $y(t,t_0,y_0)$ is the solution of (11.20)
through (t_0,y_0). This proves that $\psi(t)$ is uniformly stable for all
$t_0 \in R$.

Now we assume that $\phi(t)$ is uniformly asymptotically stable.

Then $\phi_k(t)$ is a uniformly asymptotically stable solution of (11.21)
with the same $(\delta_0, \varepsilon, T(\varepsilon))$ as the one for $\phi(t)$.   As was seen above,
$\psi(t)$ is uniformly stable with $(\varepsilon, \delta^*(\varepsilon))$, where $\delta^*(\varepsilon) = \frac{1}{2}\delta(\frac{\varepsilon}{2})$.
For a fixed ε_0 such that $0 < \varepsilon_0 < B^* - B$, set $\delta_0^* = \delta^*(\varepsilon_0)$.   For a
fixed $t_0 \in R$, if k is sufficiently large, we have
$|\phi_k(t_0) - \psi(t_0)| < \frac{1}{2}\delta_0$, where δ_0 is the one for the uniformly asymptotic stability of $\phi(t)$ and we can assume that $\frac{1}{2}\delta_0 \leq \delta_0^*$.   Let y_0
be such that $|\psi(t_0) - y_0| < \frac{1}{2}\delta_0$ and let $x(t)$ be the solution of
(11.20) such that $x(t_0 + \tau_k) = y_0$.   Then $x_k(t) = x(t + \tau_k)$ is the solu-
tion of (11.21) and $x_k(t_0) = y_0$.   Since $|\phi_k(t_0) - y_0| < \delta_0$ and
$\phi_k(t)$ is uniformly asymptotically stable, we have

$$|\phi_k(t) - x_k(t)| < \frac{\varepsilon}{2} \quad \text{for} \quad t \geq t_0 + T(\frac{\varepsilon}{2}).$$

The sequence $\{x_k(t)\}$ converges to the solution $y(t)$ of (11.20)
through (t_0, y_0) uniformly on any compact interval $t_0 + T(\frac{\varepsilon}{2}) \leq$
$t \leq t_0 + T(\frac{\varepsilon}{2}) + N$, and hence, if k is sufficiently large,

$$|x_k(t) - y(t)| < \frac{\varepsilon}{4} \quad \text{and} \quad |\phi_k(t) - \psi(t)| < \frac{\varepsilon}{4}$$

on $t_0 + T(\frac{\varepsilon}{2}) \leq t \leq t_0 + T(\frac{\varepsilon}{2}) + N$, which implies that

$$|y(t) - \psi(t)| < \varepsilon \quad \text{on} \quad t_0 + T(\frac{\varepsilon}{2}) \leq t \leq t_0 + T(\frac{\varepsilon}{2}) + N.$$

Since N is arbitrary, we have

$$|\psi(t) - y(t, t_0, y_0)| < \varepsilon \quad \text{for all} \quad t \geq t_0 + T(\frac{\varepsilon}{2})$$

if $|\psi(t_0) - y_0| < \frac{1}{2}\delta_0$.   This proves that $\psi(t)$ is uniformly asymp-
totically stable for all $t_0 \in R$.

As was seen in Theorem 11.4, if $f(t,x)$ is uniformly Lipschitzian in x, that is, satisfies (11.16) and if the zero solution
of (11.3) is uniformly asymptotically stable, there exists a Liapunov
function $V(t,x)$ defined on $0 \leq t < \infty$, $|x| < H_1$ which satisfies

the condition that

      (i) $a(|x|) \leq V(t,x) \leq b(|x|)$, where $a(r)$, $b(r)$ are continuous, increasing, positive definite,

     (ii) $|V(t,x)-V(t,y)| \leq K|x-y|$ for some constant $K > 0$,

 (iii) $\dot{V}_{(11.3)}(t,x) \leq -V(t,x)$.

This <u>Liapunov function</u> implies the <u>integrally asymptotic stability</u> of the zero solution of (11.3) and the converse also holds, see [70].   Chow and Yorke [11] have shown that this is also equivalent to saying that $x \equiv 0$ is a solution and is unique in the future and is integrally attracting for (11.3), by constructing a Liapunov function in a simpler method than [70].   Following their paper, we shall discuss a <u>converse theorem on integrally asymptotic stability</u>.   The existence as a solution of the zero function can be characterized in terms of Liapunov functions [35].   Consider a system

$$x' = f(t,x),\tag{11.26}$$

where $f(t,x)$ is continuous on $I \times S_c$, $S_c = \{x; |x| < c\}$, and $f(t,0) \equiv 0$, and consider its perturbed system

$$y' = f(t,y) + p(t),\tag{11.27}$$

where $p(t)$ is a continuous function on I.

    <u>Definition 11.1</u>.   The zero solution of (11.26) is <u>integrally stable</u>, if for any $\varepsilon > 0$, any $t_0 \geq 0$ and any $p(t)$, there exists a $\delta(\varepsilon) > 0$ such that $|y_0| < \delta(\varepsilon)$ and $\int_{t_0}^{\infty} |p(t)|dt < \delta(\varepsilon)$ imply $|y(t,t_0,y_0)| < \varepsilon$ for all $t \geq t_0$, where $y(t,t_0,y_0)$ is a solution of (11.27).

    <u>Definition 11.2</u>.   The zero solution of (11.26) is <u>integrally</u>

attracting, if there exists a $\delta_0 > 0$ and for any $\varepsilon > 0$, any $t_0 \geq 0$
and any $p(t)$, there exists a $T(\varepsilon) > 0$ and an $\eta(\varepsilon) > 0$ such that
if $|y_0| < \delta_0$ and $\int_{t_0}^{\infty} |p(t)| dt < \eta(\varepsilon)$, then $|y(t,t_0,y_0)| < \varepsilon$ for

all $t \geq t_0 + T(\varepsilon)$, where $y(t,t_0,y_0)$ is a solution of (11.27).

Definition 11.3. The zero solution of (11.26) is integrally
asymptotically stable, if it is integrally stable and is integrally
attracting.

We shall denote a solution of (11.26) through (t_0,x_0) by
$x(t,t_0,x_0)$ and a solution of (11.27) through (t_0,y_0) by $y(t,t_0,y_0)$.
Let $0 < a < c$ and let $S_a = \{x; \ |x| < a\}$. For each $t \in (0,\infty)$ and
$x \in S_a$, $A_a(t,x)$ will denote the set of absolutely continuous functions $\phi: I \to R^n$ which satisfy

$$\phi(0) = 0, \ \phi(t) = x \quad \text{and} \quad \sup_{s \in [0,t]} |\phi(s)| \leq a.$$

Let $V(t,x)$ be defined by

$$V(t,x) = \begin{cases} \displaystyle\inf_{\phi \in A_a(t,x)} \int_0^t e^{-\lambda(t-u)} |\phi'(u) - f(u;\phi(u))| du, & t > 0 \\[2mm] |x|, & t = 0, \end{cases} \tag{11.28}$$

where $\lambda \geq 0$ is a constant.

Lemma 11.3. For $\tau > 0$ and $\xi \in S_a$, there exists a solution
$x(t)$ of (11.26) such that $x(0) = 0$, $x(\tau) = \xi$ and $|x(t)| \leq a$ for
$0 \leq t \leq \tau$ if and only if $V(\tau,\xi) = 0$.

Proof. It is clear that the condition is necessary. Now suppose that $V(\tau,\xi) = 0$. Then there exists a sequence of absolutely
continuous functions $\{x_k(t)\}$, $x_k(t) \in A_a(\tau,\xi)$, such that

$$\lim_{k \to \infty} \int_0^\tau e^{-\lambda(\tau-u)} |x_k'(u) - f(u,x_k(u))| du = 0.$$

Since we have

$$e^{-\lambda\tau}\int_0^\tau |x_k'(u)-f(u,x_k(u))|\,du \leq \int_0^\tau e^{-\lambda(\tau-u)}|x_k'(u)-f(u,x_k(u))|\,du,$$

we have

$$\lim_{k\to\infty}\int_0^\tau |x_k'(u)-f(u,x_k(u))|\,du = 0. \qquad (11.29)$$

If we set

$$\phi_k(t) = x_k(t) - \int_0^t f(u,x_k(u))\,du, \quad 0 \leq t \leq \tau,$$

clearly, by (11.29), $\lim_{k\to\infty}\phi_k(t) = 0$. For $t_1,t_2, 0 \leq t_1 \leq t_2 \leq \tau$, letting $z_k(t) = x_k(t) - \phi_k(t)$, we have

$$z_k(t_2) - z_k(t_1) = \int_{t_1}^{t_2} f(u,x_k(u))\,du,$$

and hence $|z_k(t_2)-z_k(t_1)| \leq M(\tau)(t_2-t_1)$, where $M(\tau) = \max\{|f(t,x)|; 0 \leq t \leq \tau, |x| \leq a\}$. Therefore $\{z_k(t)\}$ is uniformly bounded and equicontinuous. By Ascoli's Theorem, there exists a uniformly convergent subsequence, which we denote by $\{z_k(t)\}$ again. Let $x(t)$ be the limit function. Then, clearly $x(0) = 0$ and $x(\tau) = \xi$ and

$$x(t) = \int_0^t f(u,x(u))\,du,$$

because $\phi_k(t) \to 0$ as $k \to \infty$ and thus $x_k(t) \to x(t)$ as $k \to \infty$. This shows the existence of a solution $x(t)$ such that $x(0) = 0$, $x(\tau) = \xi$ and $|x(t)| \leq a$.

Lemma 11.4. For any $t \geq s > 0$ and $x,y \varepsilon S_a$,

$$|V(s,x)-V(t,y)| \leq |x-y| + |s-t|M(t)+(1-e^{-\lambda(t-s)})a. \qquad (11.30)$$

Proof. It is sufficient to prove that

$$|V(s,x)-V(s,y)| \leq |x-y| \qquad (11.31)$$

and

$$|V(s,y)-V(t,y)| \leq |s-t|M(t)+(1-e^{-\lambda(t-s)})a. \qquad (11.32)$$

For $\phi \varepsilon A_a(s,x)$ and $0 < h < s$, let $\phi_h \varepsilon A_a(s,y)$ be a function such that $\phi_h = \phi$ on $[0,s-h]$ and the graph of ϕ_h on $[s-h,s]$ is a straight line between $(s-h,\phi(s-h))$ and (s,y). Then for all $h > 0$

$$V(s,y) \leq \int_0^s e^{-\lambda(s-u)}|\phi_h'(u)-f(u,\phi_h(u))|du$$

$$\leq \int_0^{s-h} e^{-\lambda(s-u)}|\phi'(u)-f(u,\phi(u))|du + \int_{s-h}^s |\phi_h'(u)|du$$

$$+ \int_{s-h}^s |f(u,\phi_h(u))|du$$

$$\leq \int_0^s e^{-\lambda(s-u)}|\phi'(u)-f(u,\phi(u))|du+|y-\phi(s-h)|+hM(s).$$

Since this is true for all $h > 0$ and since this is true for all $\phi \varepsilon A_a(s,x)$, letting $h \to 0$, we have

$$V(s,y) \leq V(s,x) + |y-x|. \qquad (11.33)$$

This inequality is symmetric in x and y, so (11.31) is proved.

Note that $V(t,0) \equiv 0$ by definition of V. Hence we have

$$0 \leq V(s,x) \leq |x|. \qquad (11.34)$$

Now let ϕ be in $A_a(t,y)$. Then we have

$$\int_0^t e^{-\lambda(t-u)}|\phi'(u)-f(u,\phi(u))|du$$

$$\geq e^{-\lambda(t-s)}\{\int_0^s e^{-\lambda(s-u)}|\phi'(u)-f(u,\phi(u))|du+\int_s^t |\phi'(u)|du$$

$$- \int_s^t |f(u,\phi(u))|du$$

$$\geq e^{-\lambda(t-s)}\{V(s,\phi(s)) + |y-\phi(s)|\} - (t-s)M(t)$$

$$\geq e^{-\lambda(t-s)}V(s,y)-(t-s)M(t) \qquad \text{(by (11.33))}.$$

Since this is true for all $\phi \in A_a(t,y)$ and since it follows from (11.34) that

$$(e^{-\lambda(t-s)}-1)V(s,y) \geq (e^{-\lambda(t-s)}-1)|y| \geq (e^{-\lambda(t-s)}-1)a,$$

we have

$$V(t,y) \geq V(s,y)-(t-s)M(t) + (e^{-\lambda(t-s)}-1)a. \qquad (11.35)$$

For any $\phi \in A_a(s,y)$, define $\phi^* \in A_a(t,y)$ by

$$\phi^* \equiv \phi \quad \text{on} \quad [0,s] \quad \text{and} \quad \phi^* \equiv y \quad \text{on} \quad (s,t].$$

Then for any $\phi \in A_a(s,y)$

$$V(t,y) \leq \int_0^t e^{-\lambda(t-u)} |\phi^{*\prime}(u)-f(u,\phi^*(u))|\,du$$

$$\leq e^{-\lambda(t-s)}\int_0^s e^{-\lambda(s-u)} |\phi^\prime(u)-f(u,\phi(u))|\,du+\int_s^t |f(u,y)|\,du.$$

Since this inequality is satisfied for all $\phi \in A_a(s,y)$, we have

$$V(t,y) \leq V(s,y) + (t-s)M(t).$$

This inequality with (11.35) implies (11.32).

Lemma 11.5. The function $V(t,x)$ is continuous on $0 \leq t < \infty$, $|x| < a$.

Proof. By Lemma 11.4, the continuity of $V(t,x)$ at (t,x), $t > 0$, is clear. Since $V(0,x) = |x|$ by the definition and $0 \leq V(t,x) \leq |x|$ by (11.34), to see that V is continuous at $(0,x)$, it is sufficient to prove

$$V(t,x) \geq e^{-\lambda t}|x|-tM(t). \qquad (11.36)$$

For $\phi \in A_a(t,x)$, we have

$$\int_0^t e^{-\lambda(t-u)} |\phi'(u)-f(u,\phi(u))| du$$

$$\geq e^{-\lambda t} \int_0^t |\phi'(u)| du - \int_0^t |f(u,\phi(u))| du$$

$$\geq e^{-\lambda t} |\int_0^t \phi'(u) du| - tM(t) = e^{-\lambda t}|x| - tM(t),$$

which implies (11.36).

Lemma 11.6. For $0 \leq t < \infty$ and $|x| < a$, we have

$$\dot{V}_{(11.26)}(t,x) \leq -\lambda V(t,x). \qquad (11.37)$$

Proof. Let ψ be a solution of (11.26) such that $\psi(t) = x$. For $\phi \in A_a(t,x)$, define $\phi_h \in A_a(t+h, \psi(t+h))$, $h > 0$, by

$$\phi_h = \begin{cases} \phi & \text{on } [0,t] \\ \psi & \text{on } [t,t+h] \end{cases}.$$

Then we have

$$V(t+h,\psi(t+h)) \leq \int_0^{t+h} e^{-\lambda(t+h-u)} |\phi_h'(u)-f(u,\phi_h(u))| du$$

$$\leq e^{-\lambda h} \int_0^t e^{-\lambda(t-u)} |\phi'(u)-f(u,\phi(u))| du$$

for all $\phi \in A_a(t,x)$, and hence

$$V(t+h,\psi(t+h)) \leq e^{-\lambda h} V(t,x),$$

which implies (11.37).

Lemma 11.7. Let $y(t)$ be an absolutely continuous function on $\alpha \leq t \leq \beta$ such that $|y(t)| \leq a < c$. Then, given $\varepsilon > 0$, there exists a continuous function $x(t)$ with its continuous derivative $x'(t)$ satisfying $x(\alpha) = y(\alpha)$, $x(\beta) = y(\beta)$ and

$$\left| \int_{\alpha}^{\beta} |x'(t)-f(t,x(t))|\, dt - \int_{\alpha}^{\beta} |y'(t)-f(t,y(t))|\, dt \right| < \varepsilon.$$

Proof. Since $f(t,x)$ is uniformly continuous on $\alpha \le t \le \beta$, $|x| \le \frac{a+c}{2}$, given $\varepsilon > 0$ there is a $\delta(\varepsilon) > 0$ such that if $|x-y| < \delta(\varepsilon)$, $\delta(\varepsilon) < \varepsilon$, then $|f(t,x)-f(t,y)| < \varepsilon/\beta-\alpha$ for all $\alpha \le t \le \beta$. Since $y'(t)$ is integrable, there exists a continuous function $u(t)$ such that

$$\int_{\alpha}^{\beta} |y'(t)-u(t)|\, dt < \frac{1}{2}\, \delta\left(\frac{\varepsilon}{2}\right), \tag{11.38}$$

where δ is the one for uniform continuity. Set $z(t) = y(\alpha) + \int_{\alpha}^{t} u(s)\, ds$ and $v(t) = ((y(\beta)-z(\beta))/(\beta-\alpha))(t-\alpha)$. Then the function $x(t) = z(t) + v(t)$ is continuous with its derivative and $x(\alpha) = y(\alpha)$, $x(\beta) = y(\beta)$. Clearly we have

$$|y(t)-z(t)| \le \int_{\alpha}^{t} |y'(s)-u(s)|\, ds < \frac{1}{2}\, \delta\left(\frac{\varepsilon}{2}\right)$$
$$\text{for } \alpha \le t \le \beta, \tag{11.39}$$

and hence $|y(t)-x(t)| \le |y(t)-z(t)|+|z(t)-x(t)| < \frac{1}{2}\, \delta\left(\frac{\varepsilon}{2}\right) + |y(\beta)-z(\beta)| < \delta\left(\frac{\varepsilon}{2}\right)$ by (11.39), which implies that $|f(t,x(t))-f(t,y(t))| < \dfrac{\varepsilon}{2(\beta-\alpha)}$ for $\alpha \le t \le \beta$.

On the other hand, we have

$$|y'(t)-x'(t)| \le |y'(t)-u(t)| + \frac{|y(\beta)-z(\beta)|}{\beta-\alpha},$$

which implies that by (11.38) and (11.39),

$$\int_{\alpha}^{\beta} |y'(t)-x'(t)|\, dt \le \frac{1}{2}\delta\left(\frac{\varepsilon}{2}\right) + \frac{1}{2}\delta\left(\frac{\varepsilon}{2}\right) = \delta\left(\frac{\varepsilon}{2}\right).$$

Thus

$$\left| \int_{\alpha}^{\beta} |x'(t)-f(t,x(t))|\, dt - \int_{\alpha}^{\beta} |y'(t)-f(t,y(t))|\, dt \right|$$

$$\leq \int_{\alpha}^{\beta} \bigl| \,|x'(t)-f(t,x(t))|-|y'(t)-f(t,y(t))| \,\bigr|\,dt$$

$$\leq \int_{\alpha}^{\beta} |x'(t)-y'(t)|\,dt + \int_{\alpha}^{\beta} |f(t,x(t))-f(t,y(t))|\,dt$$

$$< \delta(\tfrac{\varepsilon}{2}) + \tfrac{\varepsilon}{2} < \varepsilon.$$

Theorem 11.6.   If the zero solution of (11.26) is integrally stable, for some a, $0 < a < c$, there exists a Liapunov function $V(t,x)$ defined on $I \times S_a$ which satisfies the following conditions;

   (i) $b(|x|) \leq V(t,x) \leq |x|$, where $b(r)$ is continuous and positive definite,

   (ii) $|V(t,x)-V(t,y)| \leq |x-y|$,

   (iii) $\overset{\bullet}{V}_{(11.26)}(t,x) \leq 0$.

Proof.   For an a, $0 < a < c$, and $\lambda = 0$, define $V(t,x)$ by (11.28).   Then, by Lemma 11.5, $V(t,x)$ is continuous on $I \times S_a$ and, by Lemma 11.4, $V(t,x)$ satisfies (ii).   Condition (iii) follows from Lemma 11.6.   Clearly integral stability implies that the zero solution of (11.26) is uniformly stable and hence it is unique to the right. Therefore $V(t,x) > 0$ if $x \neq 0$ and clearly $V(t,0) \equiv 0$.   Thus we only have to prove that $V(t,x)$ is positive definite.   Suppose not. Then there exists an $\varepsilon > 0$, $0 < \varepsilon < a$, and sequences $\{t_k\}$ and $\{x_k\}$ such that $\varepsilon \leq |x_k| < a$ and

$$t_k \to \infty, \; V(t_k,x_k) \to 0 \;\; \text{as} \;\; k \to \infty.$$

Let $\delta(\varepsilon)$ be the number in Definition 11.1.   Choose a k so large that $V(t_k,x_k) < \delta(\varepsilon)$ and let $\phi_k \in A_a(t_k,x_k)$ be chosen such that

$$\int_0^{t_k} |\phi_k'(u)-f(u,\phi_k(u))|\,du < \delta(\varepsilon).$$

By Lemma 11.7, there exists a continuous function $x(t)$ with continuous derivative such that

$$\int_0^{t_k} |x'(u)-f(u,x(u))| du < \delta(\epsilon) \tag{11.40}$$

and

$$x(0) = 0, \quad x(t_k) = x_k.$$

Define

$$p(t) = \begin{cases} x'(t)-f(t,x(t)) & \text{for } t \epsilon [0,t_k] \\ 0 & \text{for } t \epsilon (t_k,\infty) \end{cases},$$

where we can assume that $p(t)$ is continuous by changing it a little and $\int_0^\infty |p(t)| dt < \delta(\epsilon)$ by (11.40). Then $x(t)$ is a solution of

$x' = f(t,x) + p(t)$ through $(0,0)$ on the interval $0 \le t \le t_k$, but $|x(t_k)| = |x_k| > \epsilon$. This contradicts the definition of integral stability. This proves the theorem.

Theorem 11.7. If the zero solution of (11.26) is unique in the future and is integrally attracting for (11.26), for some a, $0 < a < c$, there exists a Liapunov function $V(t,x)$ defined on $I \times S_a$ which satisfies the conditions (i), (ii) in Theorem 11.6 and

(iii)' $\dot{V}_{(11.26)}(t,x) \le -V(t,x)$.

Proof. Let δ_0 correspond to the δ_0 in Definition 11.2. For $\delta_0^* < \delta_0$, let $a = \delta_0^*$. For $\lambda = 1$, define $V(t,x)$ by (11.28). It is sufficient to prove the positive definiteness of $V(t,x)$. Suppose not. Then there exists an ϵ, $0 < \epsilon < \delta_0^*$, and sequences $\{t_k\}$ and $\{x_k\}$ such that $\epsilon \le |x_k| < \delta_0^*$ and

$$t_k \to \infty, \quad V(t_k,x_k) \to 0 \quad \text{as } k \to \infty.$$

Let $T(\epsilon)$ and $\eta(\epsilon)$ be numbers corresponding to those in Definition 11.2.

Let k be sufficiently large so that $t_k > T(\varepsilon) + 1$ and $V(t_k, x_k) < \eta(\varepsilon)e^{-(T(\varepsilon)+1)}$ and let $\phi_k \varepsilon A_a(t_k, x_k)$ be chosen such that

$$\int_0^{t_k} e^{-(t_k-u)} |\phi_k'(u) - f(u, \phi_k(u))| du < \eta(\varepsilon)e^{-(T(\varepsilon)+1)}.$$

Set $t_k - (T(\varepsilon)+1) = t_0$. Then $t_0 \geq 0$ and $t_k > t_0 + T(\varepsilon)$. Clearly

$$\int_{t_0}^{t_k} e^{-(t_k-u)} |\phi_k'(u) - f(u, \phi_k(u))| du < \eta(\varepsilon)e^{-(T(\varepsilon)+1)}$$

and

$$e^{-(T(\varepsilon)+1)} \int_{t_0}^{t_k} |\phi_k'(u) - f(u, \phi_k(u))| du = \int_{t_0}^{t_k} e^{-(t_k-t_0)} |\phi_k'(u) -$$

$$- f(u, \phi_k(u))| du$$

$$\leq \int_{t_0}^{t_k} e^{-(t_k-u)} |\phi_k'(u) - f(u, \phi_k(u))| du$$

$$< \eta(\varepsilon)e^{-(T(\varepsilon)+1)},$$

and hence, we have

$$\int_{t_0}^{t_k} |\phi_k'(u) - f(u, \phi_k(u))| du < \eta(\varepsilon).$$

By Lemma 11.7, there exists a continuous function $x(t)$ with continuous derivative such that

$$x(t_0) = \phi_k(t_0), \quad x(t_k) = x_k$$

and

$$\int_{t_0}^{t_k} |x'(u) - f(u, x(u))| du < \eta(\varepsilon).$$

Define

$$p(t) = \begin{cases} x'(t) - f(t, x(t)) & \text{for } t \varepsilon [t_0, t_k] \\ 0 & \text{for } t \varepsilon (t_k, \infty), \end{cases}$$

where we can assume that $p(t)$ is continuous and $\int_{t_0}^{\infty} |p(t)| dt < \eta(\varepsilon)$.

Then $x(t)$ is a solution of $x' = f(t,x) + p(t)$ on $t_0 \leq t \leq t_k$

such that $|x(t_0)| = |\phi_k(t_0)| < \delta_0$. However $|x(t_k)| = |x_k| > \varepsilon$,
which contradicts the definition of integral attraction since
$t_k > t_0 + T(\varepsilon)$. This proves the theorem.

Remark 1. If the zero solution of (11.26) is integrally
asymptotically stable, system (11.26) can be perturbed by a larger
class of functions, that is, <u>interval bounded functions</u>. For the details, see [11].

Remark 2. The zero solution of

$$x' = \begin{cases} x \sin^2 \dfrac{1}{x} & (x \neq 0) \\[2mm] 0 & (x = 0) \end{cases}$$

is uniformly stable, but not integrally stable.

12. Total Stability

Consider a system

$$x' = f(t,x), \qquad\qquad (12.1)$$

where $f(t,x) \in C(I \times S_{B*}, R^n)$, $S_{B*} = \{x;\ |x| < B*\}$.

Definition 12.1. Let $\phi(t)$ be a solution of (12.1) which
satisfies $|\phi(t)| \le B$, $B < B*$, for all $t \ge 0$. The solution $\phi(t)$
is said to be <u>totally stable</u>, if for any $t_0 \ge 0$ and any $\varepsilon > 0$
there exists a $\delta(\varepsilon) > 0$ such that if $g(t,x)$ is any continuous
function on $[t_0, \infty) \times S_{B*}$ and satisfies

$$|g(t,x) - f(t,x)| < \delta(\varepsilon)$$

for all $(t,x) \in [t_0, \infty) \times S_{B*}$, $|\phi(t) - x| \le \varepsilon$, and if $x_0 \in S_{B*}$ satis-
fies $|\phi(t_0) - x_0| < \delta(\varepsilon)$, then any solution $x(t)$ through (t_0, x_0)
of the system

$$x' = g(t,x) \qquad\qquad (12.2)$$

satisfies $|\phi(t)-x(t)| < \varepsilon$ for all $t \geq t_0$.

Let $\phi(t)$ be the solution of (12.1) in Definition 12.1. Then we have the following equivalence.

Lemma 12.1. $\phi(t)$ is totally stable if and only if for any $t_0 \geq 0$ and any $\varepsilon > 0$ there exists a $\delta(\varepsilon) > 0$ such that if $h(t)$ is any continuous function on $[t_0,\infty)$ and satisfies $|h(t)| < \delta(\varepsilon)$ for all $t \geq t_0$ and if $y_0 \varepsilon S_{B*}$ satisfies $|\phi(t_0)-y_0| < \delta(\varepsilon)$, then any solution $y(t)$ through (t_0,y_0) of the system

$$y' = f(t,y) + h(t) \qquad\qquad (12.3)$$

satisfies $|\phi(t)-y(t)| < \varepsilon$ for all $t \geq t_0$.

Proof. The necessity of the condition is clear. Suppose that there exists a $t_0 \varepsilon I$, a function $g(t,x)$ and a solution $x(t)$ of (12.2) such that $|\phi(t_0)-x(t_0)| < \delta(\varepsilon)$ and $|x(t_1)-\phi(t_1)| = \varepsilon$ for some $t_1 > t_0$, although $|g(t,x)-f(t,x)| < \delta(\varepsilon)$ for all $(t,x) \varepsilon$ $[t_0,\infty) \times S_{B*}$, $|\phi(t)-x| \leq \varepsilon$, where $\delta(\varepsilon)$ is the one given in the condition. Here we can assume that $\varepsilon < B*-B$ and $|\phi(t)-x(t)| \leq \varepsilon$ for all $t,t_1 \geq t \geq t_0$. If we set

$$h(t) = g(t,x(t)) - f(t,x(t)) \quad \text{for} \quad t_1 \geq t \geq t_0,$$

$|h(t)| < \delta(\varepsilon)$ for $t_1 \geq t \geq t_0$ and $h(t)$ is continuous on $[t_0,t_1]$. Moreover, $h(t)$ can be easily extended to the interval $[t_0,\infty)$ so that $|h(t)| < \delta(\varepsilon)$ for all $t \geq t_0$. Now we consider the system (12.3) with this function $h(t)$. Then we can find a solution $y(t)$ of (12.3) such that $y(t) = x(t)$ for $t \leq t_1$. Obviously, $|\phi(t_0)-y(t_0)| < \delta(\varepsilon)$ and $|\phi(t_1)-y(t_1)| = \varepsilon$. Thus there arises a contradiction. This proves the lemma.

Now consider the case where $f(t,x)$ of (12.1) satisfies $f(t,0) \equiv 0$ and

$$|f(t,x)-f(t,y)| \leq L|x-y| \quad \text{for } t \in I, \ x \in S_{B*}, \ y \in S_{B*}.$$

In this case, if the zero solution of (12.1) is uniformly asymptotically stable, as was seen in Theorem 11.4, there exists a Liapunov function $V(t,x)$ defined on $0 \leq t < \infty$, $|x| < H_1$, where H_1 is a suitable constant, which satisfies the condition that

(i) $a(|x|) \leq V(t,x) \leq b(|x|)$, where $a(r)$, $b(r)$ are continuous, increasing, positive definite,

(ii) $|V(t,x)-V(t,y)| \leq K|x-y|$ for some constant $K > 0$,

(iii) $\dot{V}_{(12.1)}(t,x) \leq -V(t,x)$.

As was seen in the previous section, the existence of such a <u>Liapunov function</u> is equivalent to the integrally asymptotic stability of the zero solution of (12.1). Therefore, assume now that there exists a Liapunov function $V(t,x)$ which satisfies the conditions above. For any $\varepsilon > 0$, choose a $\delta_1 = \delta_1(\varepsilon) > 0$ so that $K\delta_1 < a(\varepsilon)$, and choose a $\delta = \delta(\varepsilon) > 0$ so that $\delta < \delta_1$ and $b(\delta) < K\delta_1$. Consider a solution $y(t,t_0,y_0)$ of (12.3), where $|h(t)| < \delta(\varepsilon)$ and $|y_0| < \delta$. Then we have

$$V'(t,y(t,t_0,y_0)) \leq -V(t,y(t,t_0,y_0)) + K\delta_1$$

as long as $y(t,t_0,y_0)$ exists. Consider a scalar linear equation

$$u' = -u + K\delta_1,$$

which has the solution $u(t)$ with initial value $V(t_0,y_0)$ at $t = t_0$, where

$$u(t) = e^{-(t-t_0)}(V(t_0,y_0)-K\delta_1) + K\delta_1.$$

Comparing $V(t,y(t,t_0,y_0))$ with $u(t)$, we have

$$V(t,y(t,t_0,y_0)) \leq K\delta_1,$$

because $V(t_0,y_0) \leq b(\delta) < K\delta_1$. On the other hand, since $K\delta_1 < a(\varepsilon)$, we have

$$a(|y(t,t_0,y_0)|) \leq V(t,y(t,t_0,y_0)) < a(\varepsilon),$$

and hence, $|y(t,t_0,y_0)| < \varepsilon$ for all $t \geq t_0$. This shows that the zero solution of (12.1) is totally stable. Thus we have the following theorem.

Theorem 12.1. Suppose that the system (12.1) has the zero solution which is integrally asymptotically stable. Then it is totally stable.

Clearly total stability implies uniform stability, but the converse is not true. For example, consider $x' = 0$. Moreover, total stability does not necessarily imply asymptotic stability even for a scalar autonomous equation. The following example shows this fact [48].

Example 12.1. Let $f(x)$ be a continuous function which vanishes at the points $\frac{1}{2^k}$ and $-\frac{1}{2^k}$, $k = 1,2,\ldots$. Moreover, assume that $xf(x) < 0$ for $x \neq 0$, $x \neq \frac{1}{2^k}$, $x \neq -\frac{1}{2^k}$. Consider an equation

$$x' = f(x). \tag{12.4}$$

Then it is clear that the zero solution of (12.4) is not asymptotically stable, because of the existence of solutions $x = \frac{1}{2^k}$ arbitrarily near $x = 0$. However, it is totally stable. In fact, given $\varepsilon > 0$ choose k so that $\frac{1}{2^k} < \varepsilon$ and let $\delta \leq \frac{1}{2^{k+1}}$ be so small that   f(x)   takes values $> \delta$ in both intervals $(\frac{1}{2^{k+1}}, \frac{1}{2^k})$ and

$(-\frac{1}{2^k}, -\frac{1}{2^{k+1}})$. Then, if the function h(t)   satisfies $|h(t)| < \delta$,

we have $x(f(x)+h(t)) < 0$ at certain points of both intervals and

any solution x(t) of $x' = f(x) + h(t)$ such that $|x(t_0)| < \delta$

cannot leave the interval $|x| \leq \frac{1}{2^k} < \varepsilon$ for $t \geq t_0$.

For a linear system

$$x' = A(t)x, \qquad\qquad (12.5)$$

where A(t) is an $n \times n$ continuous matrix on I, if the zero solu-

tion is uniformly asymptotically stable, that is, exponentially

asymptotically stable, there exists a Liapunov function V(t,x) which

satisfies the conditions in Theorem 11.1. Therefore the zero solution

is totally stable. The following theorem is the reciprocal [47].

Theorem 12.2. If the zero solution of (12.5) is totally stable,

then it is uniformly asymptotically stable.

Proof. If the zero solution is totally stable, there exists a

$\delta > 0$ such that if $|y_0| < \delta$, the solution $y(t,t_0,y_0)$ of

$$y' = A(t)y + \delta y,$$

where $|y| < 1$, satisfies $|y(t,t_0,y_0)| < 1$. But the solutions of

both equations are related by

$$y(t,t_0,y_0) = x(t,t_0,y_0)e^{\delta(t-t_0)}.$$

Thus we have $|x(t,t_0,y_0)| < e^{-\delta(t-t_0)}$, $|y_0| < \delta$, which proves the

uniformly asymptotic stability of the zero solution of (12.5).

We shall now consider an almost periodic system

$$x' = f(t,x), \qquad\qquad (12.6)$$

where $f(t,x) \in C(R \times S_{B*}, R^n)$ and $f(t,x)$ is almost periodic in t

uniformly for $x \in S_{B*}$. Let K be a given compact set in S_{B*}, and

let $\phi(t)$ be a solution of (12.6) such that $\phi(t) \in K$ for all $t \geq 0$.

For $g \in H(f)$ and $p \in H(f)$, define $\rho(g,p,K)$ by

$$\rho(g,p,K) = \sup\{|g(t,x)-p(t,x)|; \quad t \in R, \ x \in K\}.$$

Sell [64] introduced the following stability which is equivalent to

the $\sum$-stability introduced by Seifert [61].

   Definition 12.2. The solution $\phi(t)$ is said to be stable under

disturbances from H(f) for $t \geq 0$ with respect to K, if for any

$\varepsilon > 0$ there exists a $\delta(\varepsilon) > 0$ such that $|\phi(t+\tau)-x(t,0,x_0,g)| \leq \varepsilon$

for $t \geq 0$, whenever $g \in H(f)$, $|\phi(\tau)-x_0| \leq \delta(\varepsilon)$ and $\rho(f_\tau,g,K) \leq \delta(\varepsilon)$

for some $\tau \geq 0$, where $f_\tau = f(t+\tau,x)$ and $x(t,0,x_0,g)$ is a solu-

tion of

$$x' = g(t,x) \tag{12.7}$$

such that $x(0,0,x_0,g) = x_0$ and $x(t,0,x_0,g) \in K$ for all $t \geq 0$.

   Remark 1. The zero solution of $x' = 0$ is not totally stable,

but clearly it is stable under disturbances from the hull.

   Remark 2. The stability under disturbances of $\phi(t)$ can be

represented in the following way. The solution $\phi(t)$ is stable under

disturbances from H(f) with respect to K, if for any $\varepsilon > 0$ and

$\tau \geq 0$, there exists a $\delta(\varepsilon) > 0$ such that $|\phi(t)-x(t,\tau,x_0,g)| \leq \varepsilon$

for $t \geq \tau$, whenever $g \in H(f)$, $|\phi(\tau)-x_0| \leq \delta(\varepsilon)$ and $\rho(f,g,K) \leq \delta(\varepsilon)$,

where $x(t,\tau,x_0,g)$ is a solution of (12.7) through (τ,x_0) and

$x(t,\tau,x_0,g) \in K$ for all $t \geq \tau$.

   Theorem 12.3. Let $\phi(t)$ be a solution of the almost periodic

system (12.6) which satisfies $|\phi(t)| \leq B, \ B < B*$, for all $t \geq 0$.

If $\phi(t)$ is totally stable for $t \geq 0$, then it is stable under dis-
turbances from $H(f)$ with respect to K, $K = \{x; \ |x| \leq B_1,$
$B < B_1 < B^*\}$.

This theorem is clear from the definition and Remark 2.

Theorem 12.4. Let $\phi(t)$ be a solution of the almost periodic
system (12.6) which satisfies $|\phi(t)| \leq B$, $B < B^*$, for all $t \geq 0$.
If $\phi(t)$ is stable under disturbances from $H(f)$ with respect to
K, $K = \{x; \ |x| \leq B_1,\ B < B_1 < B^*\}$, then $\phi(t)$ is asymptotically almost periodic in t.

Proof. Let $\{\tau_k\}$ be any sequence such that $\tau_k \to \infty$ as
$k \to \infty$. Set $\phi_k(t) = \phi(t+\tau_k)$. It is sufficient to show that the
sequence $\{\phi_k(t)\}$ has a subsequence which converges uniformly on
$0 \leq t < \infty$. Clearly $\phi_k(t)$ is a solution of

$$x' = f(t + \tau_k, x) \qquad\qquad (12.8)$$

through $(0,\phi(\tau_k))$ and $|\phi_k(t)| \leq B$ for all $t \geq 0$. It is clear
that $\phi_k(t)$ is stable under disturbances from $H(f_{\tau_k})$ with respect
to K with the same pair $(\varepsilon,\delta(\varepsilon))$ as the one for $\phi(t)$. Since K
is compact, the sequence $\{\tau_k\}$ has a subsequence, which we shall de-
note by $\{\tau_k\}$ again, such that $f(t+\tau_k, x)$ converges uniformly on
$R \times K$ as $k \to \infty$, and hence there is an integer $k_0(\varepsilon) > 0$ such that
if $m \geq k \geq k_0(\varepsilon)$,

$$|f(t+\tau_k,x) - f(t+\tau_m,x)| \leq \delta(\varepsilon) \quad \text{on} \quad R \times K, \qquad (12.9)$$

where $\delta(\varepsilon)$ is the one for the stability under disturbances. Therefore, if $m \geq k \geq k_0(\varepsilon)$

$$\rho(f_{\tau_k}, f_{\tau_m}, K) \leq \delta(\varepsilon).$$

Furthermore, since $|\phi_k(0)| \leq B$, we can assume that if $m \geq k \geq k_0(\varepsilon)$,

$$\left| \phi_k(0) - \phi_m(0) \right| \le \delta(\varepsilon), \qquad (12.10)$$

taking a subsequence again, if necessary.

Since $\phi_m(t)$ is a solution of

$$x' = f(t + \tau_m, x) \qquad (12.11)$$

and $\left| \phi_m(t) \right| \le B$ for all $t \ge 0$ and since $f_{\tau_m} \in H(f_{\tau_k}) = H(f)$ and $\phi_k(t)$ is stable under disturbances from $H(f_{\tau_k})$ with respect to K, we have

$$\left| \phi_k(t) - \phi_m(t) \right| \le \varepsilon \quad \text{for all} \quad t \ge 0$$

if $m \ge k \ge k_0(\varepsilon)$. This proves that $\phi(t + \tau_k)$ is uniformly convergent on I as $k \to \infty$. Thus $\phi(t)$ is asymptotically almost periodic.

Corollary 12.1. Let $\phi(t)$ be a solution of the almost periodic system (12.6) which satisfies $\left| \phi(t) \right| \le B$, $B < B^*$, for all $t \ge 0$. If $\phi(t)$ is totally stable, then $\phi(t)$ is asymptotically almost periodic in t.

This follows immediately from Theorems 12.3 and 12.4.

Now we shall see the relationship between stability under disturbances and uniform stability in a periodic system

$$x' = f(t,x), \quad f(t+\omega,x) = f(t,x), \quad \omega > 0, \qquad (12.12)$$

where $f(t,x) \in C(R \times S_{B^*}, R^n)$.

In the case where f is not autonomous on $R \times S_{B^*}$, there is a smallest positive period ω^* of $f(t,x)$ on $R \times S_{B^*}$ and we can see that for any $g \in H(f)$ and any $\tau \ge 0$ there is a $\sigma(\tau,g)$ such that

$$\tau - \frac{\omega^*}{2} \le \sigma(\tau,g) \le \tau + \frac{\omega^*}{2}$$

and

$$g(t,x) = f(t+\sigma,x) \quad \text{on} \quad R \times S_{B*}.$$

For such a $\sigma(\tau,g)$, we have the following lemma.

Lemma 12.2. For any $\varepsilon > 0$, there exists a $\gamma(\varepsilon) > 0$ such that if $\tau \geq 0$, $g \in H(f)$ and $\rho(f_\tau,g) \leq \gamma(\varepsilon)$, then $|\tau - \sigma(\tau,g)| < \varepsilon$.

Proof. Suppose that there is no $\gamma(\varepsilon)$. Then there is an $\varepsilon > 0$ and there are sequences $\{\gamma_k\}$, $\{\tau_k\}$ and $\{\sigma_k\}$ such that

$$\gamma_k \to 0 \quad \text{as} \quad k \to \infty, \quad \tau_k \geq 0,$$

$$\sup\{|f(t+\tau_k,x)-f(t+\sigma_k,x)|; \ t \in R, \ x \in S_{B*}\} < \gamma_k,$$

$$\tau_k - \frac{\omega^*}{2} \leq \sigma_k \leq \tau_k + \frac{\omega^*}{2}$$

and

$$|\tau_k - \sigma_k| \geq \varepsilon.$$

Set $\tau_k = N_k\omega^* + \tau_k'$, where N_k is a nonnegative integer and $0 \leq \tau_k' < \omega^*$. If we set $\sigma_k = N_k\omega^* + \sigma_k'$, then

$$\tau_k' - \frac{\omega^*}{2} \leq \sigma_k' \leq \tau_k' + \frac{\omega^*}{2}.$$

Since $0 \leq \tau_k' < \omega^*$ and $-\frac{\omega^*}{2} \leq \sigma_k' \leq \omega^* + \frac{\omega^*}{2}$, there are τ' and σ' such that $\tau_k' \to \tau'$, $\sigma_k' \to \sigma'$ as $k \to \infty$, taking a subsequence, if necessary, Then $\tau' - \frac{\omega^*}{2} \leq \sigma' \leq \tau' + \frac{\omega^*}{2}$, that is,

$$|\tau' - \sigma'| \leq \frac{\omega^*}{2}. \tag{12.13}$$

On the other hand, $\gamma_k \to 0$ as $k \to \infty$ and

$$\sup\{|f(t+\tau_k,x)-f(t+\sigma_k,x)|; \ t \in R, \ x \in S_{B*}\}$$

$$= \sup\{|f(t+\tau_k',x)-f(t+\sigma_k',x)|; \ t \in R, \ x \in S_{B*}\} < \gamma_k,$$

and hence we have

$$f(t+\tau',x) = f(t+\sigma',x) \quad \text{on} \quad R \times S_{B*}.$$

This shows that $|\tau'-\sigma'|$ is a period of $f(t,x)$ on $R \times S_{B^*}$. Since $|\tau_k'-\sigma_k'| = |\tau_k-\sigma_k| \geq \varepsilon$ implies $\varepsilon \leq |\tau'-\sigma'|$ and since we also have (12.13), this contradicts that ω^* is the smallest positive period of $f(t,x)$. This proves the lemma.

Theorem 12.5. Let $\phi(t)$ be a solution of the periodic system (12.12) such that $|\phi(t)| \leq B$, $B < B^*$, for all $t \geq 0$. If $\phi(t)$ is uniformly stable, then $\phi(t)$ is stable under disturbances from $H(f)$.

Proof. If f is autonomous on S_{B^*} , that is, $f(t,x) = f^*(x)$ on $R \times S_{B^*}$, then for any $g \in H(f)$, $g(t,x) = f^*(x)$. Therefore $\rho(f_\tau,g) = 0$. Thus it is clear that $\phi(t)$ is stable under disturbances from $H(f)$ since $\phi(t)$ is uniformly stable.

We shall now consider the case where f is not autonomous and we assume ω^* to be the smallest positive period of $f(t,x)$. For $t,t' \in [0,\infty)$, we have

$$|\phi(t)-\phi(t')| < \frac{\delta(\varepsilon)}{2} \quad \text{if} \quad |t-t'| < \frac{\delta(\varepsilon)}{2L} ,$$

where $\delta(\varepsilon)$ is the number for the uniform stability of $\phi(t)$ and $L > 0$ is such that $|f(t,x)| \leq L$ for $t \in R$, $|x| \leq \frac{B^*+B}{2}$. By Lemma 12.2, there is a $\gamma(\varepsilon) > 0$ such that $\tau \geq 0$, $g \in H(f)$ and $\rho(f_\tau,g) \leq \gamma(\varepsilon)$ imply $|\tau-\sigma(\tau,g)| < \frac{\delta(\varepsilon)}{2L}$, where we can assume that $\gamma(\varepsilon) < \frac{\delta(\varepsilon)}{2}$ and $0 < \varepsilon < \frac{B^*-B}{2}$. Moreover,

$$g(t,x) = f(t+\sigma,x) \quad \text{on} \quad R \times S_{B^*} .$$

For a fixed $\tau \geq 0$, let $\psi(t) = \phi(t+\tau)$. Then $\psi(t)$ is a solution through $(0,\phi(\tau))$ of

$$x' = f(t+\tau,x). \tag{12.14}$$

Let y_0 be such that $|\phi(\tau)-y_0| \leq \gamma(\varepsilon)$ and let $g \in H(f)$ be such that $\rho(f_\tau, g) \leq \gamma(\varepsilon)$. Consider a solution $x(t)$ through $(0, y_0)$ of

$$x' = g(t,x). \tag{12.15}$$

As long as $x(t)$ exists, $x(t)$ is a solution of

$$x' = f(t+\sigma, x), \quad \sigma = \sigma(\tau, g), \tag{12.16}$$

through $(0, y_0)$. Hence we have

$$|\tau - \sigma| < \frac{\delta(\varepsilon)}{2L}. \tag{12.17}$$

If we set $y(t) = \psi(t+\sigma-\tau)$, then $y(t) = \phi(t+\sigma)$. First of all, we assume that $\sigma \geq 0$. Then $y(t)$ is a solution of (12.16) through
$(0, \phi(\sigma))$ and $y(t)$ is uniformly stable for $t \geq 0$ with the same
pair $(\varepsilon, \delta(\varepsilon))$ as the one for $\phi(t)$. (12.17) implies
$|\phi(\tau) - \phi(\sigma)| < \frac{\delta(\varepsilon)}{2}$ and

$$|y(0) - y_0| = |\phi(\sigma) - y_0| \leq |\phi(\sigma) - \phi(\tau)| + |\phi(\tau) - y_0|$$

$$< \frac{\delta(\varepsilon)}{2} + \gamma(\varepsilon)$$

$$< \delta(\varepsilon),$$

and hence the uniform stability of $y(t)$ implies that

$$|y(t) - x(t)| < \varepsilon \quad \text{for all} \quad t \geq 0. \tag{12.18}$$

Moreover, since we have (12.17),

$$|y(t) - \psi(t)| = |\psi(t+\sigma-\tau) - \psi(t)| < \frac{\delta(\varepsilon)}{2} \quad \text{for all} \quad t \geq 0,$$

and hence $|\psi(t) - x(t)| < 2\varepsilon$ or $|\phi(t+\tau) - x(t)| < 2\varepsilon$ for all $t \geq 0$.

Next we shall consider the case where $\sigma < 0$, and consequently,
$\tau - \sigma > 0$. If we set $z(t) = x(t+\tau-\sigma)$, then $z(t)$ is a solution of
(12.18) through $(0, x(\tau-\sigma))$. Since (12.17) implies $|y_0 - x(\tau-\sigma)| <$
$\frac{\delta(\varepsilon)}{2}$, we have

$$|\psi(0)-z(0)| = |\phi(\tau)-x(\tau-\sigma)| \leq |\phi(\tau)-y_0| + |y_0-x(\tau-\sigma)| < \delta(\varepsilon).$$

Thus we have $|\psi(t)-z(t)| < \varepsilon$ for all $t \geq 0$, because $\psi(t)$ is uni-formly stable. Moreover,

$$|z(t)-x(t)| = |x(t+\tau-\sigma)-x(t)| < \frac{\delta(\varepsilon)}{2} < \varepsilon \text{ for all } t \geq 0,$$

and therefore $|\psi(t)-x(t)| < 2\varepsilon$ or $|\phi(t+\tau)-x(t)| < 2\varepsilon$ for all $t \geq 0$. Thus we see that $\phi(t)$ is stable under disturbances from $H(f)$.

Corollary 12.2. Let $\phi(t)$ be a solution of the periodic sys-tem (12.16) such that $|\phi(t)| \leq B$, $B < B^*$, for all $t \geq 0$. If $\phi(t)$ is uniformly stable, then $\phi(t)$ is asymptotically almost periodic in t.

This corollary follows immediately from Theorem 12.4 and 12.5.

All results, except Theorem 12.1, in this section hold also for <u>functional differential equations</u> $\dot{x}(t) = f(t,x_t)$ under the as-sumption that $|f(t,\phi)| \leq L(\alpha)$, $t \in R$, $|\phi| \leq \alpha$. For the details, see [34], [36], [84], [85].

13.   <u>Inherited Properties in Almost Periodic Systems</u>
       Consider an almost periodic system

$$x' = f(t,x), \qquad\qquad (13.1)$$

where $f(t,x) \in C(R \times S_{B^*}, R^n)$, $S_{B^*} = \{x; |x| < B^*\}$, and $f(t,x)$ is almost periodic in t uniformly for $x \in S_{B^*}$. In this section, let $\phi(t)$ be a solution of (13.1) which is defined on $t \geq 0$ and satis-fies $|\phi(t)| \leq B$ for all $t \geq 0$, where $B < B^*$. For some sequence $\{\tau_k\}$ such that $\tau_k > 0$, let $f(t+\tau_k,x) \to g(t,x)$ uniformly on $R \times S$, S any compact set in S_{B^*}, as $k \to \infty$. Then $g \in H(f)$. Moreover, assume that $\phi(t+\tau_k) \to \psi(t)$ uniformly on any compact set in I as $k \to \infty$. Clearly $\psi(t)$ is a solution of

$$x' = g(t,x), \quad g \in H(f). \tag{13.2}$$

We shall denote by $(\psi,g) \in H(\phi,f)$ this fact.

Definition 13.1. A property P is said to be <u>inherited</u> if when ϕ has the property P with respect to the solutions of (13.1), ψ also has the property P with respect to the solutions of (13.2).

This definition was formally given by Fink [18].

Let K be the set of x such that $|x| \leq B_1$, $B < B_1 < B^*$. The following theorem shows that total stability and stability under disturbances are inherited properties.

Theorem 13.1. Let $\{\tau_k\}$ be a sequence such that $\tau_k > 0$, $\phi(\tau_k) \to x_0$ as $k \to \infty$ and that $f(t+\tau_k,x) \to g(t,x)$ uniformly on $R \times K$ as $k \to \infty$. Then, if $\phi(t)$ is totally stable, the solution $\psi(t)$ of (13.2) through $(0,x_0)$ is totally stable. Moreover, if $\phi(t)$ is stable under disturbances from $H(f)$ with respect to K, then $\psi(t)$ is stable under disturbances from $H(g)$ with respect to K.

Proof. Now assume that $\phi(t)$ is totally stable. Then $\phi_k(t) = \phi(t+\tau_k)$ is a solution of

$$x' = f(t+\tau_k,x) \tag{13.3}$$

and is totally stable with the same pair $(\varepsilon,\delta(\varepsilon))$ as the one for $\phi(t)$. By the assumption, it is easily seen that a subsequence of $\phi_k(t)$ converges to a solution $\eta(t)$ of (13.2) through $(0,x_0)$ uniformly on any compact interval of I. We shall denote by $\{\phi_k(t)\}$ the subsequence again.

For an $\varepsilon > 0$ and $t_0 \geq 0$, let $y(t,t_0,y_0,h)$ be a solution of

$$x' = g(t,x) + h(t) \qquad\qquad (13.4)$$

such that $|\eta(t_0)-y_0| < \frac{1}{2}\,\delta(\frac{\varepsilon}{2})$, where $|h(t)| < \frac{1}{2}\,\delta(\frac{\varepsilon}{2})$ for $t \geq t_0$. If k is sufficiently large, we have

$$|\phi_k(t_0)-\eta(t_0)| < \frac{1}{2}\,\delta(\frac{\varepsilon}{2}) \quad\text{and}\quad |f(t+\tau_k,x)-g(t,x)| < \frac{1}{2}\,\delta(\frac{\varepsilon}{2})$$
for $(t,x) \in R \times K$.

Since $\phi_k(t)$ is totally stable, this implies that

$$|\phi_k(t)-\eta(t)| < \frac{\varepsilon}{2} \quad\text{for}\quad t \geq t_0.$$

On the other hand, $|\phi_k(t_0)-y_0| < \delta(\frac{\varepsilon}{2})$ and

$$|f(t+\tau_k,x)-g(t,x)-h(t)| \leq \delta(\frac{\varepsilon}{2}),$$

which imply that $|\phi_k(t)-y(t,t_0,y_0,h)| < \frac{\varepsilon}{2}$ for $t \geq t_0$. Thus we have

$$|\eta(t)-y(t,t_0,y_0,h)| < \varepsilon \quad\text{for}\quad t \geq t_0$$

if $|\eta(t_0)-y_0| < \frac{1}{2}\,\delta(\frac{\varepsilon}{2})$ and $|h(t)| < \frac{1}{2}\,\delta(\frac{\varepsilon}{2})$ for $t \geq t_0$. This proves that $\eta(t)$ is totally stable. Stability implies the uniqueness, and hence $\psi(t)$ is totally stable.

Now assume that $\phi(t)$ is stable under disturbances from $H(f)$ with respect to K. Clearly $\phi_k(t)$ is stable under disturbances from $H(f_{\tau_k})$ with respect to K with the same $(\varepsilon,\delta(\varepsilon))$. For an $\varepsilon > 0$ and $\tau \geq 0$, let $y(t,0,y_0,h)$ be a solution of

$$x' = h(t,x) \qquad\qquad (13.5)$$

such that $|\eta(\tau)-y_0| \leq \frac{1}{2}\,\delta(\frac{\varepsilon}{2})$, where $\eta(t)$ is the same solution as before and $\rho(g_\tau,h,K) \leq \frac{1}{2}\,\delta(\frac{\varepsilon}{2})$. If k is sufficiently large, $|g(t+\tau,x)-f(t+\tau_k+\tau,x)| \leq \frac{1}{2}\,\delta(\frac{\varepsilon}{2})$ for all $t \in R$ and $x \in K$ and $|\eta(\tau)-\phi_k(\tau)| \leq \frac{1}{2}\,\delta(\frac{\varepsilon}{2})$. Therefore $\rho(f_{\tau_k+\tau},h,K) \leq \delta(\frac{\varepsilon}{2})$ and

$|\phi_k(\tau)-y_0| \leq \delta(\frac{\varepsilon}{2})$, which imply that

$$|\phi_k(t+\tau)-y(t,0,y_0,h)| < \frac{\varepsilon}{2} \quad \text{for} \quad t \geq 0.$$

However, if k is sufficiently large, $|\phi_k(t+\tau)-\eta(t+\tau)| < \frac{\varepsilon}{2}$ on $[0,N]$, and hence

$$|\eta(t+\tau)-y(t,0,y_0,h)| < \varepsilon \quad \text{for all} \quad t \geq 0$$

if $|\eta(\tau)-y_0| \leq \frac{1}{2}\delta(\frac{\varepsilon}{2})$ and $\rho(g_\tau,h,K) \leq \frac{1}{2}\delta(\frac{\varepsilon}{2})$, because N is arbitrary. This proves that the solution $\psi(t)$ is stable under disturbances from $H(g)$ with respect to K.

Corollary 13.1. Under the assumptions in Theorem 13.1, if $\phi(t)$ is totally stable, then the solution $\psi(t)$ must be uniformly stable. Moreover, if $\phi(t)$ is stable under disturbances from $H(f)$ with respect to K, then the solution $\psi(t)$ must be uniformly stable.

Now consider a periodic system

$$x' = f(t,x), \quad f(t+\omega,x) = f(t,x), \quad \omega > 0, \tag{13.6}$$

where $f(t,x) \varepsilon C(R \times S_{B*},R^n)$. The following theorem shows that uniform stability is an inherited property and uniformly asymptotic stability is also inherited.

Theorem 13.2. Let $\phi(t)$ be a solution of the periodic system (13.6) such that $|\phi(t)| \leq B$, $B < B*$, for all $t \geq 0$. Let $\{\tau_k\}$ be a sequence such that $\tau_k > 0$, $\phi(\tau_k) \to x_0$ and $f(t+\tau_k,x) \to g(t,x)$ as $k \to \infty$. If $\phi(t)$ is uniformly stable, then the solution $\psi(t)$ through $(0,x_0)$ of

$$x' = g(t,x) \tag{13.7}$$

is also uniformly stable. Moreover, if $\phi(t)$ is uniformly asymptotically stable, then $\psi(t)$ is uniformly asymptotically stable.

$\underline{Proof}$. Since $f(t,x)$ is periodic in t, there is a subsequence $\{\tau_{k_j}\}$ of $\{\tau_k\}$ such that

$$f(t+\tau_{k_j},x) \to g(t,x) \quad \text{uniformly on} \quad R \times \overline{S}_{B_1}, \quad B < B_1 < B^*,$$

as $k \to \infty$.

Set $\phi_k(t) = \phi(t+\tau_k)$. By Corollary 12.2, $\phi(t)$ is asymptotically almost periodic, and hence we can assume that $\phi_{k_j}(t)$ converges uniformly on I. Thus there is a solution $\eta(t)$ of (13.7) which passes through $(0,x_0)$ and for which $\phi_{k_j}(t) \to \eta(t)$ uniformly on I as $j \to \infty$.

Now we shall see that $\eta(t)$ is uniformly stable. Then $\eta(t)$ will be a unique solution of (13.7) through $(0,x_0)$, and hence $\eta(t) \equiv \psi(t)$. Set $\tau_k = N_k\omega + \sigma_k$, where $N_k \geq 0$ is an integer such that $N_k\omega \leq \tau_k < (N_k+1)\omega$. Here we can assume that $\sigma_{k_j} \to \sigma$ as $j \to \infty$. Then $0 \leq \sigma \leq \omega$ and $g(t,x) = f(t+\sigma,x)$. For any $\varepsilon > 0$, $\varepsilon < \dfrac{B_1 - B}{2}$, let $\delta(\varepsilon)$ be the one for the uniform stability of $\phi(t)$. For any $t_0 \varepsilon I$, let $x(t)$ be a solution of (13.7) such that $|\eta(t_0)-x(t_0)| < \delta(\varepsilon)$. For a fixed $x(t)$, we have $|\eta(t_0)-x(t_0)| = \gamma < \delta(\varepsilon)$. If j is sufficiently large,

$$|\phi_{k_j}(t_0)-\eta(t_0)| < \frac{\delta(\varepsilon)-\gamma}{2}$$

and

$$|\phi(t_0+\sigma+N_{k_j}\omega)-\phi(t_0+\sigma_{k_j}+N_{k_j}\omega)| < \frac{\delta(\varepsilon)-\gamma}{2},$$

and hence we have

$$|\phi(t_0+\sigma+N_{k_j}\omega) - x(t_0)| < \delta(\varepsilon).$$

Since $\phi(t+\sigma+N_{k_j}\omega)$ is a uniformly stable solution of (13.7) where

$g(t,x) = f(t+\tau,x)$, we have

$$|\phi(t+\sigma+N_{k_j}\omega)-x(t)| < \varepsilon \quad \text{for all} \quad t \geq t_0. \tag{13.8}$$

On the other hand, for an arbitrary $\rho > 0$, if j is sufficiently large,

$$|\eta(t_0)-\phi(t_0+\sigma+N_{k_j}\omega)| \leq |\eta(t_0)-\phi(t_0+\sigma_{k_j}+N_{k_j}\omega)|$$

$$+ |\phi(t_0+\sigma_{k_j}+N_{k_j}\omega) - \phi(t_0+\sigma+N_{k_j}\omega)| < \delta(\rho),$$

and hence $|\eta(t)-\phi(t+\sigma+N_{k_j}\omega)| < \rho$ for all $t \geq t_0$. From this and (13.8), it follows that

$$|\eta(t)-x(t)| < \varepsilon + \rho \quad \text{for all} \quad t \geq t_0.$$

Since ρ is arbitrary, we have

$$|\eta(t)-x(t)| \leq \varepsilon \quad \text{for all} \quad t \geq t_0 \quad \text{if} \quad |\eta(t_0)-x(t_0)| < \delta(\varepsilon).$$

This proves that $\eta(t)$ is uniformly stable.

Next we assume that $\phi(t)$ is uniformly asymptotically stable. Let $x(t)$ be a solution of (13.7) such that $|\eta(t_0)-x(t_0)| < \delta_0$, where δ_0 is the number in the definition of uniformly asymptotic stability. For a fixed $x(t)$, $|\eta(t_0)-x(t_0)| = \delta_1 < \delta_0$ and if j is sufficiently large,

$$|\phi(t_0+\sigma+N_{k_j}\omega)-x(t_0)| \leq |\phi(t_0+\sigma+N_{k_j}\omega)-\phi(t_0+\sigma_{k_j}+N_{k_j}\omega)|$$

$$+ |\phi(t_0+\sigma_{k_j}+N_{k_j}\omega)-\eta(t_0)| + |\eta(t_0)-x(t_0)| < \delta_0.$$

Since $\phi(t+\sigma+N_{k_j}\omega)$ is a uniformly asymptotically stable solution of (13.7) with the same δ_0 as the one for $\phi(t)$, we have

$$|\phi(t+\sigma+N_{k_j}\omega)-x(t)| < \varepsilon \quad \text{for} \quad t \geq t_0 + T(\varepsilon)$$

if j is sufficiently large. Moreover, if j is sufficiently large,

$$|\phi(t_0+\sigma+N_{k_j}\omega)-\eta(t_0)| \le |\phi(t_0+\sigma+N_{k_j}\omega)-\phi(t_0+\sigma_{k_j}+N_{k_j}\omega)|$$

$$+ |\phi(t_0+\sigma_{k_j}+N_{k_j}\omega)-\eta(t_0)| < \delta_0,$$

and hence, for sufficiently large, j, $|\phi(t+\sigma+N_{k_j}\omega)-\eta(t)| < \varepsilon$ for all $t \ge t_0 + T(\varepsilon)$.   Thus we have

$$|\eta(t)-x(t)| < 2\varepsilon \quad \text{for} \quad t \ge t_0+T(\varepsilon) \quad \text{if} \quad |\eta(t_0)-x(t_0)| < \delta_0.$$

This shows that $\eta(t)$ is uniformly asymptotically stable.   This completes the proof.

For the almost periodic system (13.1), uniform stability and uniformly asymptotic stability are not necessarily inherited without assuming the uniqueness.   The following example due to Kato [34] shows this fact.   This example is very important because it tells us many things.

Now let $a_0(t) \equiv 1$ and let $a_k(t)$ be a periodic function with period 2^k such that

$$a_k(t) = \begin{cases} 0 & (0 \le t < 2^{k-1}) \\ -\dfrac{1}{2^k} & (2^{k-1} \le t < 2^k). \end{cases}$$

Making $a_k(t)$ smooth, define $a(t)$ by $a(t) = \sum_{k=0}^{\infty} a_k(t)$.   Then $a(t)$ is an almost periodic function and clearly $a(t) > 0$.   Let $h(x)$ be such that

$$h(x) = \begin{cases} 0 & (x = 0) \\ 2\sqrt{n|x-1/n|} & (\dfrac{2}{2n+1} \le x \le \dfrac{2}{2n-1}) \end{cases}$$

and define $f(t,x)$ for $t \in R$ and $0 \le x \le 2$ by

$$f(t,x) = h(x)-ca(t)\sqrt{x}, \quad c > 2\sqrt{2}.$$

Consider an almost periodic equation

$$x' = \begin{cases} f(t,x), & t \in R, \quad 0 \le x \le 2 \\ -f(t,-x), & t \in R, \quad -2 \le x < 0. \end{cases} \qquad (13.9)$$

The zero solution of (13.9) is uniformly asymptotically stable. This will be proved in the following way. Since we have

$$f(t,\frac{1}{n}) = h(\frac{1}{n}) - ca(t)\sqrt{\frac{1}{n}} = -ca(t)\sqrt{\frac{1}{n}} < 0,$$

the solution of (13.9) cannot cross $x = \frac{1}{n}$ from below to above. This implies the uniform stability of the zero solution. Since $a_0(t) \equiv 1$ and $a_1(t) = 0$ on the interval $2k < t < 2k+1$, we have

$$a(t) \ge \frac{1}{2} \quad \text{on} \quad 2k < t < 2k+1.$$

Moreover, since $h(x) \le \sqrt{2x}$ for $x \ge 0$, we have

$$f(t,x) \le \sqrt{2}\,\sqrt{x} - \frac{c}{2}\,\sqrt{x} = \frac{1}{2}(2\sqrt{2} - c)\,\sqrt{x}$$

for $2k < t < 2k+1$ and $x \ge 0$. Comparing with the solution of $x' = \frac{1}{2}(2\sqrt{2}-c)\,\sqrt{x}$, if a solution $x(t)$ of (13.9) satisfies

$$\frac{1}{n-1} \ge x(t) \ge \frac{1}{n} \quad \text{on} \quad [a,b] \subset (2k,2k+1),$$

we have

$$b-a < \frac{4}{c-2\sqrt{2}}\left(\frac{1}{\sqrt{n-1}} - \frac{1}{\sqrt{n}}\right) = T_n.$$

Since $T_n \to 0$ as $n \to \infty$, we can see that the zero solution of (13.9) is uniformly asymptotically stable.

Now consider a sequence $\{2^k-1\}$. Then $f(t+2^k-1,x) \to h(x)$ on $0 \le t \le 1$. Let $g(t,x)$ be a function in $H(f)$ for the sequence $\{2^k-1\}$. Then the zero solution of

$$x' = g(t,x) \qquad (13.10)$$

is not unique to the right. This will be shown in the following way.

On the interval $0 \le t \le 1$, $g(t,x) = h(x)$. Consider the equation

$$x' = h(x) \tag{13.11}$$

on $0 \le t \le 1$. As long as a solution $x(t)$ ($\ne \frac{1}{n}$) of (13.11) through $(t_n, \frac{1}{n})$ satisfies $\frac{1}{n} \le x(t) \le \frac{2}{2n-1}$, we have

$$x(t) = \frac{1}{n} + n(t-t_n)^2 ,$$

and as long as a solution $x(t)$ ($\ne \frac{1}{n-1}$) of (13.11) through $(t_{n-1}, \frac{1}{n-1})$ satisfies $\frac{2}{2n-1} \le x(t) \le \frac{1}{n-1}$, we have

$$x(t) = \frac{1}{n-1} - (n-1)(t-t_{n-1})^2$$

Choose $\{t_n\}$ so that

$$t_{n-1}-t_n = (\frac{1}{n} + \frac{1}{n-1}) \frac{1}{\sqrt{2n-1}} .$$

Then we obtain a solution $x(t)$ through $(t_n, \frac{1}{n})$ which goes to the left. Clearly

$$t_n = t_1 - \sum_{k=1}^{n-1} (\frac{1}{k+1} + \frac{1}{k}) \frac{1}{\sqrt{2k+1}}$$

and

$$\sum_{k=1}^{\infty} (\frac{1}{k+1} + \frac{1}{k}) \frac{1}{\sqrt{2k+1}} \le \sum_{k=1}^{\infty} \frac{1}{\sqrt{2} \ k^{3/2}} ,$$

which is convergent, and this implies that $x(t)$ reaches $x = 0$ in a finite time $t = t_\infty$. Namely, the zero solution is not unique to the right, and consequently it is not stable.

For the almost periodic system (13.1), we have the following theorem.

Theorem 13.3. Suppose that for every $g \ \varepsilon \ H(f)$, the solution of (13.2) is unique for the initial value problem. Let $\{\tau_k\}$ be a sequence such that $\tau_k > 0$, $f(t+\tau_k,x) \to g(t,x)$ uniformly on $R \times K$

and $\phi(\tau_k) \to x_0$ as $k \to \infty$. If $\phi(t)$ is uniformly stable, then the

solution $\psi(t)$ of (13.2) through $(0,x_0)$ is uniformly stable, More-

over, if $\phi(t)$ is uniformly asymptotically stable, then $\psi(t)$ is

also uniformly asymptotically stable.

    Proof. Clearly, $\phi_k(t) = \phi(t+\tau_k)$ is a uniformly stable solu-

tion of

$$x' = f(t+\tau_k,x) \tag{13.12}$$

through $(0,\phi(\tau_k))$ with the same pair $(\varepsilon,\delta(\varepsilon))$ as the one for

$\phi(t)$, and $|\phi_k(t)| \leq B$ for $t \geq 0$. Therefore, $\{\phi_k(t)\}$ is uniformly

bounded and equicontinuous on I, and hence there exists a subsequence,

which we shall denote by $\{\phi_k(t)\}$ again, such that $\phi_k(t)$ converges

to $\psi(t)$ uniformly on any compact interval on I. For a fixed

$t_0 \in I$, if k is sufficiently large, we have

$$\left|\phi_k(t_0)-\psi(t_0)\right| < \tfrac{1}{2} \delta(\tfrac{\varepsilon}{2}), \tag{13.13}$$

where we can assume that $\varepsilon < B^*-B$. Let y_0 be such that

$$\left|y_0-\psi(t_0)\right| < \tfrac{1}{2} \delta(\tfrac{\varepsilon}{2}). \tag{13.14}$$

and let $x(t)$ be the solution of (13.1) such that $x(t_0+\tau_k) = y_0$.

Then $x_k(t) = x(t+\tau_k)$ is a solution of (13.12) and $x_k(t_0) = y_0$.

Since $\phi_k(t)$ is uniformly stable and $|\phi_k(t_0)-y_0| < \delta(\tfrac{\varepsilon}{2})$ by (13.13)

and (13.14), we have

$$\left|\phi_k(t)-x_k(t)\right| < \tfrac{\varepsilon}{2} \quad \text{for all} \quad t \geq t_0. \tag{13.15}$$

Since $|x_k(t)| \leq B + \tfrac{\varepsilon}{2}$ for all $t \geq t_0$, the sequence $\{x_k(t)\}$ con-

verges to the solution $y(t)$ of (13.2) through (t_0,y_0), which is

uniquely determined, uniformly on any compact interval $[t_0,t_0+N]$.

Thus, if k is sufficiently large,

$$\left| x_k(t) - y(t) \right| < \frac{\varepsilon}{4} \quad \text{and} \quad \left| \phi_k(t) - \psi(t) \right| < \frac{\varepsilon}{4} \quad \text{on} \ [t_0, t_0+N]. \quad (13.16)$$

It follows from (13.15) and (13.16) that $\left| \psi(t) - y(t) \right| < \varepsilon$ on $[t_0, t_0+N]$. Since N is arbitrary, $\left| \psi(t) - y(t, t_0, y_0) \right| < \varepsilon$ for all $t \geq t_0$ if $\left| \psi(t_0) - y_0 \right| < \frac{1}{2} \delta(\frac{\varepsilon}{2})$, where $y(t, t_0, y_0)$ is the solution of (13.2) through (t_0, y_0). This proves that $\psi(t)$ is uniformly stable.

Now we assume that $\phi(t)$ is uniformly asymptotically stable. Then $\phi_k(t)$ is a uniformly asymptotically stable solution of (13.12) with the same $(\delta_0, \varepsilon, T(\varepsilon))$ as the one for $\phi(t)$. As was seen above, $\psi(t)$ is uniformly stable with $(\varepsilon, \delta^*(\varepsilon))$, where $\delta^*(\varepsilon) = \frac{1}{2} \delta(\frac{\varepsilon}{2})$. For a fixed ε_0 such that $0 < \varepsilon_0 < B^* - B$, let $\delta_0^* = \delta^*(\varepsilon_0)$. For a fixed $t_0 \varepsilon I$, if k is sufficiently large, we have $\left| \phi_k(t_0) - \psi(t_0) \right| < \frac{1}{2} \delta_0$, where δ_0 is the number in the definition of uniformly asymptotic stability and we can assume that $\frac{1}{2} \delta_0 \leq \delta_0^*$.

Let y_0 be such that $\left| \psi(t_0) - y_0 \right| < \frac{1}{2} \delta_0$ and let $x(t)$ be the solution of (13.1) such that $x(t_0 + \tau_k) = y_0$. Then $x_k(t) = x(t + \tau_k)$ is the solution of (13.12) through (t_0, y_0). Since $\left| \phi_k(t_0) - y_0 \right| < \delta_0$ and $\phi_k(t)$ is uniformly asymptotically stable,

$$\left| \phi_k(t) - x_k(t) \right| < \frac{\varepsilon}{2} \quad \text{for} \ t \geq t_0 + T(\frac{\varepsilon}{2}).$$

The sequence $\{x_k(t)\}$ converges to the solution $y(t)$ of (13.2) through (t_0, y_0) uniformly on any compact interval $t_0 + T(\frac{\varepsilon}{2}) \leq t \leq t_0 + T(\frac{\varepsilon}{2}) + N$, and hence, if k is sufficiently large,

$$\left| x_k(t) - y(t) \right| < \frac{\varepsilon}{4} \quad \text{and} \quad \left| \phi_k(t) - \psi(t) \right| < \frac{\varepsilon}{4} \quad \text{on}$$

$$t_0 + T(\frac{\varepsilon}{2}) \leq t \leq t_0 + T(\frac{\varepsilon}{2}) + N.$$

Therefore $\left| y(t) - \psi(t) \right| < \varepsilon$ on $t_0 + T(\frac{\varepsilon}{2}) \leq t \leq t_0 + T(\frac{\varepsilon}{2}) + N$. Since N is arbitrary, we have

$$\left| \psi(t) - y(t) \right| < \varepsilon \quad \text{for all} \ t \geq t_0 + T(\frac{\varepsilon}{2})$$

if $|\psi(t_0)-y_0| < \frac{1}{2}\delta_0$. This completes the proof.

In the above example, the zero solution of (13.9) is unique to the right, but the zero solution of (13.10) is not unique. Thus the uniqueness is not necessarily inherited. This fact will be characterized by a Liapunov Function [52].

Theorem 13.4.   For any $(\psi,g) \in H(\phi,f)$, ψ is unique to the right if and only if for any $T > 0$ and for any $\tau \geq 0$, there exists a Liapunov function $V(t,x,\tau)$ defined on $\tau \leq t \leq \tau+T$, $|\phi(t)-x| < \alpha$, $\alpha = \frac{B^*-B}{2}$, which satisfies the following conditions;

   (i) $a(|\phi(t)-x|) \leq V(t,x,\tau) \leq b(|\phi(t)-x|)$, where $a(r)$ and $b(r)$ are continuous and positive definite (independent of τ),

   (ii) $|V(t,x,\tau)-V(t,y,\tau)| \leq K|x-y|$, where $K > 0$ is a constant (independent of τ),

   (iii) $\overset{\cdot}{V}_{(13.1)}(t,x) \leq 0$.

Proof.   Let $(\psi,g) \in H(\phi,f)$. Then there exists a sequence $\{\tau_k\}$, $\tau_k \geq 0$, such that $\phi(t+\tau_k) \to \psi(t)$ and $f(t+\tau_k,x) \to g(t,x)$ uniformly on any compact subset of $I \times S_{B^*}$ as $k \to \infty$. $\phi_k(t) = \phi(t+\tau_k)$ is a solution of (13.12) through $(0,\phi(\tau_k))$. We shall show that $\psi(t)$ is the unique solution through $(0,\psi(0))$ under the assumption that there exists the Liapunov function above.

Suppose that $y(t)$ is a solution of (13.2) through $(0,\psi(0))$ such that

$$\psi(t) = y(t) \quad \text{on} \quad 0 \leq t \leq t_0,$$
$$\psi(t) \neq y(t) \quad \text{on} \quad t_0 < t \leq t_0+\gamma,$$

where we can assume that $\gamma < T$ and $t_0 = 0$ by considering $\{\tau_k+t_0\}$ instead of $\{\tau_k\}$.

If k is sufficiently large and γ is small,

$$|\psi(t)-\phi_k(t)| < \alpha, \quad |y(t)-\phi_k(t)| < \alpha \quad \text{on} \quad [0,\gamma],$$

because $\phi_k(t)$ converges to $\psi(t)$ uniformly on $[0,T]$. Set $W(t,x,\tau_k) = V(t+\tau_k,x,\tau_k)$. Then $W(t,x,\tau_k)$ is defined on $0 \leq t \leq T$, $|\phi_k(t)-x| < \alpha$. Since $W(t,x,\tau_k)$ satisfies a Lipschitz condition with respect to x, we have

$$\dot{W}_{(13.2)}(t,x,\tau_k) \leq \dot{W}_{(13.12)}(t,x,\tau_k) + K|g(t,x)-f(t+\tau_k,x)|$$

$$\leq \dot{V}_{(13.1)}(t+\tau_k,x,\tau_k) + K|g(t,x)-f(t+\tau_k,x)|$$

$$\leq K|g(t,x) -f(t+\tau_k,x)|,$$

because $\dot{V}_{(13.1)}(t+\tau_k,x,\tau_k) \leq 0$. If we let δ_k be such that

$$\delta_k = \sup\{|g(t,x)-f(t+\tau_k,x)|; \ 0 \leq t \leq T, \ |\phi_k(t)-x| < \alpha\},$$

we have $\dot{W}_{(13.2)}(t,x,\tau_k) \leq K\delta_k$ on $0 \leq t \leq T$, which implies that

$$W(t,y(t),\tau_k) \leq W(0,y(0),\tau_k) + K\delta_k t \quad \text{on} \quad [0,\gamma].$$

By condition (i), we have

$$a(|y(t)-\phi_k(t)|) \leq b(|\psi(0)-\phi_k(0)|) + K\delta_k t \quad \text{for} \quad t \ \varepsilon \ [0,\gamma].$$

Since $\delta_k \to 0$ as $k \to \infty$, we have $a(|y(t)-\psi(t)|) \leq 0$ for $t \ \varepsilon \ [0,\gamma]$, which implies that $y(t)-\psi(t) = 0$ on $[0,\gamma]$. This proves the uniqueness of $\psi(t)$. Here, note that for sufficiency, a fixed $T > 0$ is enough.

Now consider a domain $\tau \leq t \leq \tau+T$, $|x-\phi(t)| < \alpha$, where T is any positive constant. Then $|x-\phi(t)| < \alpha$ implies $|x| < B^*$. Let $A(t,x,\tau)$ be the family of absolutely continuous functions $\xi(u)$ on $[\tau,t]$, $t \leq \tau+T$, such that $\xi(\tau) = \phi(\tau)$, $\xi(t) = x$ and $|\xi(u)-\phi(u)| \leq \alpha$ for $u \ \varepsilon \ [\tau,t]$. Define $V(t,x,\tau)$ by

$$V(t,x,\tau) = \begin{cases} \inf\limits_{\xi \in A(t,x,\tau)} \int_{\tau}^{t} |\xi'(u) - f(u,\xi(u))| \, du & \text{for } t > \tau \\ & \\ |x - \phi(\tau)| & \text{for } t = \tau. \end{cases} \quad (13.17)$$

Then, as was seen in Section 11, $V(t,x,\tau)$ is continuous in (t,x) and $V(t,\phi(t),\tau) \equiv 0$. Moreover, we have

$$V(t,x,\tau) \leq |x - \phi(t)| \quad \text{for } \tau \leq t \leq \tau + T, \ |x - \phi(t)| \leq \alpha,$$

$$|V(t,x,\tau) - V(t,y,\tau)| \leq |x - y|,$$

$$\dot{V}_{(13.1)}(t,x,\tau) \leq 0.$$

By the uniqueness of the solution $\phi(t)$, $V(t,x,\tau) > 0$ if $|x - \phi(t)| > 0$. Therefore we can find an $a(r)$ in condition (i), but it may depend on τ generally.

Now let $0 < r < \alpha$ and set

$$a(r,\tau) = \inf\{V(t,x,\tau); \ \tau \leq t \leq \tau + T, \ |x - \phi(t)| = r\}.$$

Then $Q(r,\tau) = \{(t,x); \ \tau \leq t \leq \tau + T, \ |x - \phi(t)| = r\}$ is a compact set, and hence there exists a $(t_0,x_0) \in Q(r,\tau)$ such that $V(t_0,x_0,\tau) = a(r,\tau)$. Clearly $a(r,\tau) > 0$ for $r > 0$ and $\tau \in I$. We shall show that

$$\inf\{a(r,\tau); \ 0 \leq \tau < \infty\} = a(r) > 0 \quad \text{for } 0 < r < \alpha.$$

Then we can see that there exists a function independent of τ which satisfies condition (i).

Suppose $\inf\{a(r_0,\tau); \ 0 \leq \tau < \infty\} = 0$ for some r_0 such that $0 < r_0 < \alpha$. Then there exists a sequence $\{\tau_k\}$, $\tau_k \geq 0$, and $(t_k,x_k) \in Q(r_0,\tau_k)$, $t_k \geq \tau_k$, such that

$$\lim_{k \to \infty} V(t_k,x_k,\tau_k) = 0.$$

If we set $t_k - \tau_k = s_k$, then $0 \leq s_k \leq T$ and

$$\lim_{k \to \infty} W(s_k, x_k, \tau_k) = \lim_{k \to \infty} V(s_k + \tau_k, x_k, \tau_k) = 0, \qquad (13.18)$$

where $W(t, x, \tau_k)$ is defined by $W(t, x, \tau_k) = V(t + \tau_k, x, \tau_k)$. Note that $W(t, x, \tau_k)$ is defined on $0 \leq t \leq T$, $|\phi_k(t) - x| < \alpha$. If we denote by $B(t, x, \tau_k)$ the family of absolutely continuous functions $\eta(u)$ on $[0, t]$, $t \leq T$, such that $\eta(0) = \phi_k(0)$, $\eta(t) = x$ and $|\eta(u) - \phi_k(u)| \leq \alpha$ for $u \varepsilon [0, t]$, we have

$$W(t, x, \tau_k) = \begin{cases} \displaystyle\inf_{\eta \varepsilon B(t, x, \tau_k)} \int_0^t |\eta'(u) - f(u + \tau_k, \eta(u))| \, du & \text{for} \quad t > 0 \\[2mm] |x - \phi_k(0)| & \text{for} \quad t = 0. \end{cases}$$
$$(13.19)$$

We shall now see that $\displaystyle\lim_{k \to \infty} s_k \equiv s_0 > 0$. Since $|f(t + \tau_k, x)| \leq L$ for some $L > 0$ if $0 \leq t \leq T$ and $|x| \leq B + \alpha$, we have

$$W(s_k, x_k, \tau_k) \geq |x_k - \phi_k(0)| - Ls_k. \qquad (13.20)$$

On the other hand, we have

$$|x_k - \phi_k(0)| \geq |x_k - \phi_k(s_k)| - |\phi_k(s_k) - \phi_k(0)|$$

and $|x_k - \phi(\tau_k + s_k)| = r_0$, where $\phi_k(t) = \phi(t + \tau_k)$ and $t_k = \tau_k + s_k$. Therefore

$$W(s_k, x_k, \tau_k) \geq r_0 - Ls_k - Ls_k$$

$$\geq r_0 - 2L\, s_k.$$

Taking a subsequence, if necessary, if $s_0 = 0$, $\lim W(s_k, x_k, \tau_k) \geq r_0$, which contradicts (13.18). Thus $s_0 \neq 0$.

By (13.18) and (13.19), there exist $\eta_k \varepsilon B(s_k, x_k, \tau_k)$ such that

$$\lim_{k \to \infty} \int_0^{s_k} |\eta_k'(u) - f(u + \tau_k, \eta_k(u))| \, du = 0.$$

Thus, by the same argument as in the proof of Lemma 11.3, we can see that there exists a function $y(t)$ and a $g \varepsilon H(f)$ such that

$$y(t) = y(0) + \int_0^t g(s,y(s))ds \quad \text{for} \quad t \in [0,s_0),$$

this shows that $y(t)$ is a solution of (13.2).

On the other hand, choosing a subsequence, we can assume that $\{\phi_k(t)\}$ converges to some function $\psi(t)$ uniformly on any compact interval in I. Then $\psi(t)$ is a solution of (13.2) and $y(0) = \psi(0)$, because $\eta_k(0) = \phi_k(0)$. We can also assume that $s_k \to s_0$ as $k \to \infty$. If t in $[0,s_0)$ is sufficiently close to s_0 and if k is sufficiently large,

$$|\eta_k(s_k) - \eta_k(t)| < \frac{r_0}{4} \quad \text{and} \quad |\phi_k(s_k) - \phi_k(t)| < \frac{r_0}{4},$$

because $\{\phi_k\}$ is equicontinuous and $y_k(t) = \eta_k(t) - z_k(t)$ is equicontinuous, where $z_k(t) = \eta_k(t) - \phi_k(0) - \int_0^t f(u+\tau_k, \eta_k(u))du$ and $z_k(t) \to 0$ as $k \to \infty$. Since $x_k = \eta_k(s_k)$, we have

$$|y(t) - \psi(t)| \geq |x_k - \phi_k(s_k)| - \{|y(t) - \eta_k(t)| + |\eta_k(t) - \eta_k(s_k)|$$

$$+ |\phi_k(s_k) - \phi_k(t)| + |\phi_k(t) - \psi(t)|\}.$$

For all k, $|x_k - \phi_k(s_k)| = r_0$, and hence we have $|y(s_0) - \psi(s_0)| \geq \frac{r_0}{2}$. Thus $\psi(t)$ is not a unique solution of (13.2), which is a contradiction. Therefore $a(r) > 0$ for $0 < r < \alpha$. This completes the proof.

The following corollary can be easily proved [52].

Corollary 13.2. Let $T > 0$ be given. For any $(\psi,g) \in H(\phi,f)$, ψ is a unique solution to the right, if and only if for any $\varepsilon > 0$ there exists a $\delta(\varepsilon) > 0$ such that $|\phi(t) - x(t,\tau,x_0)| < \varepsilon$ on $\tau \leq t \leq \tau+T$ whenever $|x_0 - \phi(\tau)| < \delta(\varepsilon)$ and $|h(t)| < \delta(\varepsilon)$, $t \geq \tau$, at some $\tau \geq 0$, where $x(t,\tau,x_0)$ is a solution through (τ,x_0) of $x' = f(t,x) + h(t)$ and $h(t)$ is continuous.

14. Uniformly Asymptotic Stability in Almost Periodic Systems.

Consider the example in Section 13 again. The zero solution of the almost periodic equation (13.9) is uniformly asymptotically stable. But the zero solution of (13.9) is neither totally stable nor stable under disturbances from the hull, because if so, the zero solution of (13.10) must be uniformly stable by Corollary 13.1, but the zero solution of (13.10) is not stable. Thus this example tells us also that uniformly asymptotic stability in an almost periodic system does not necessarily imply total stability while it does for a periodic system, which will be seen later. In this section, we shall obtain conditions under which uniformly asymptotic stability implies total stability.

Consider an almost periodic system

$$x' = f(t,x), \qquad\qquad (14.1)$$

where $f \in C(R \times S_{B*}, R^n)$, $S_{B*} = \{x;\ |x| < B*\}$, and $f(t,x)$ is almost periodic in t uniformly for $x \in S_{B*}$. Throughout this section, we assume that $\phi(t)$ is a solution of the system defined on I such that $|\phi(t)| \leq B$, $B < B*$, for all $t \geq 0$. We shall use the same notation as in Section 13.

Lemma 14.1. We assume that for each $(\psi,g) \in H(\phi,f)$, ψ is unique for the initial value problem. Let $T > 0$ be given. Then, for any $\varepsilon > 0$ there is a $\delta(\varepsilon) > 0$ such that $t_0 \in I$, $|x_0 - \phi(t_0)| < \delta(\varepsilon)$ and $|h(t)| < \delta(\varepsilon)$ imply that

$$|\phi(t) - x(t,t_0,x_0)| < \varepsilon \quad \text{on} \quad t_0 \leq t \leq t_0 + T,$$

where $x(t,t_0,x_0)$ is a solution through (t_0,x_0) of the system

$$x' = f(t,x) + h(t) \qquad\qquad (14.2)$$

and h(t) is continuous on I.

  Proof. For ordinary differential equations, this lemma is
exactly Corollary 13.2. But we shall show the proof which does work
for functional differential equations. Now suppose that for some
$\epsilon > 0$ there is no δ which satisfies the condition in the lemma.
Then there exist sequences $\{\delta_k\}$, $\{t_k\}$, $\{\tau_k\}$ and $\{h_k(t)\}$ such that
$\delta_k \to 0$ as $k \to 0$, $t_k \in (\tau_k, \tau_k + T)$, $|h_k(t)| < \delta_k$, $|\phi(\tau_k) - x_k(\tau_k)| < \delta_k$
and

$$|\phi(t_k) - x_k(t_k)| = \frac{\epsilon}{2}, \quad |\phi(t) - x_k(t)| < \frac{\epsilon}{2} \quad \text{on} \quad \tau_k \leq t < t_k,$$

where $x_k(t)$ is a solution of the system

$$x' = f(t,x) + h_k(t) \qquad\qquad (14.3)$$

and we can assume that $\epsilon < B^* - B$. If we set $\phi_k(t) = \phi(t + \tau_k)$ and
$y_k(t) = x_k(t + \tau_k)$, then $\phi_k(t)$ is a solution of

$$x' = f(t + \tau_k, x) \qquad\qquad (14.4)$$

and $y_k(t)$ is a solution through $(0, x_k(\tau_k))$ of

$$x' = f(t + \tau_k, x) + h_k(t + \tau_k). \qquad\qquad (14.5)$$

Set $t_k = \tau_k + \sigma_k$. Then $0 < \sigma_k \leq T$, where we can assume $\sigma_k \to \sigma$,
$0 < \sigma \leq T$, as $k \to \infty$. If k is sufficiently large, $|\sigma_k - \sigma| < \frac{\epsilon}{8L}$,
where L is such that $|f(t,x)| \leq L$ for $t \in R$ and $|x| \leq B + \epsilon$.
In case $\sigma_k \geq \sigma$, clearly $y_k(t)$ exists on $[0, \sigma]$. For $\sigma_k < \sigma$, as
long as $|y_k(t)| \leq B + \epsilon$,

$$y_k(t) = y_k(\sigma_k) + \int_{\sigma_k}^{t} f(s + \tau_k, y_k(s)) ds + \int_{\sigma_k}^{t} h_k(s + \tau_k) ds, \quad t \geq \sigma_k.$$

Thus, as long as $|y_k(t)| \leq B + \epsilon$ for $\sigma_k \leq t \leq \sigma$,

$$|y_k(t) - y_k(\sigma_k)| < L \frac{\epsilon}{8L} + \delta_k \frac{\epsilon}{8L} = \frac{\epsilon}{8} + \delta_k \frac{\epsilon}{8L}.$$

Since $\left| y_k(\sigma_k) - \phi_k(\sigma_k) \right| = \left| x_k(\tau_k + \sigma_k) - \phi(\tau_k + \sigma_k) \right| = \left| x_k(t_k) - \phi(t_k) \right| = \frac{\varepsilon}{2}$,
we have

$$\left| y_k(t) - \phi_k(t) \right| \leq \left| y_k(t) - y_k(\sigma_k) \right| + \left| y_k(\sigma_k) - \phi_k(\sigma_k) \right| + \left| \phi_k(\sigma_k) - \phi_k(t) \right|$$

$$< \frac{\varepsilon}{8} + \delta_k \frac{\varepsilon}{8L} + \frac{\varepsilon}{2} + \frac{\varepsilon}{8} .$$

Thus, if k is sufficiently large, we have $\left| y_k(t) - \phi_k(t) \right| < \varepsilon$ on
$[0,\sigma]$ as long as $\left| y_k(t) \right| \leq B+\varepsilon$, and hence, if k is sufficiently
large, $y_k(t)$ exists on $[0,\sigma]$ and

$$y_k(t) = x_k(\tau_k) + \int_0^t f(s+\tau_k, y_k(s))ds + \int_0^t h_k(s+\tau_k)ds,$$

$$0 \leq t \leq \sigma. \tag{14.6}$$

There is a subsequence of $\{\tau_k\}$, which we shall denote by
$\{\tau_k\}$ again, such that $f(t+\tau_k, x)$ converges to some $g \in H(f)$ uni-
formly on $R \times S$, $S = \{x; \left| x \right| \leq B+\varepsilon\}$, and $\phi(\tau_k) \to x_0$ as $k \to \infty$.
Since $y_k(0) = x_k(\tau_k)$ and $\left| \phi(\tau_k) - x_k(\tau_k) \right| < \delta_k$, clearly $y_k(0) \to x_0$
as $k \to \infty$. Since $\{y_k(t)\}$ is uniformly bounded and equicontinuous on
$0 \leq t \leq \sigma$, there is a subsequence, which will be denoted by $\{y_k(t)\}$
again, such that $y_k(t) \to x(t)$ uniformly on $[0,\sigma]$ as $k \to \infty$. Letting $k \to \infty$, it follows from (14.6) that

$$x(t) = x_0 + \int_0^t g(s, x(s))ds, \quad 0 \leq t \leq \sigma. \tag{14.7}$$

This shows that $x(t)$ is the solution of

$$x' = g(t,x) \tag{14.8}$$

through $(0,x_0)$. On the other hand, $\phi_k(t)$ tends to the solution
$\psi(t)$ of (14.8) which also passes through $(0,x_0)$. By the uniqueness,
$x(t) \equiv \psi(t)$ on $0 \leq t \leq \sigma$. However, $\left| x_k(t_k) - \phi(t_k) \right| =$
$\left| y_k(\sigma_k) - \phi_k(\sigma_k) \right| = \frac{\varepsilon}{2}$ implies $\left| x(\sigma) - \psi(\sigma) \right| = \frac{\varepsilon}{2}$. This contradiction
proves the lemma.

Corollary 14.1. For any $(\psi,g) \in H(\phi,f)$, if ψ is uniformly stable, then the conclusion of Lemma 14.1 holds. In particular, for a periodic system

$$x' = f(t,x), \quad f(t+\omega,x) = f(t,x), \quad \omega > 0, \qquad (14.9)$$

if $\phi(t)$ is uniformly stable, the conclusion of Lemma 14.1 holds.

Note that for the periodic system, the uniform stability of $\phi(t)$ implies the uniform stability of $\psi(t)$.

By the same idea as in the proof of Lemma 14.1, we can prove the following lemma, see [36].

Lemma 14.2. Suppose that for every $g \in H(f)$ solutions of

$$x' = g(t,x) \qquad (14.10)$$

are unique for the initial value problem. Let $T > 0$ and $B_1 (< B^*)$ be given. Then, for any $\varepsilon > 0$ there exists a $\delta(\varepsilon) > 0$ such that for any $t_0 \in I$, if $x(t)$ is a solution of the system (14.1) which satisfies $|x(t)| \leq B_1$ for all $t_0 \leq t \leq t_0+T$ and if $h(t)$ is a continuous function such that $|h(t)| < \delta(\varepsilon)$ on $[t_0,t_0+T]$, we have

$$|x(t)-y(t)| < \varepsilon \quad \text{for all} \quad t_0 \leq t \leq t_0+T,$$

whenever $y(t)$ is a solution of (14.2) satisfying $|x(t_0)-y(t_0)| < \delta(\varepsilon)$.

If for any $(\psi,g) \in H(\phi,f)$, ψ is uniformly stable with a common $\delta(\cdot)$, we can say a little more. For a periodic system, if $\phi(t)$ is uniformly stable, then this condition is satisfied as the proof of Theorem 13.2 shows. As will be seen from the proof of Theorem 13.3, if we assume that for every $g \in H(f)$, solutions are unique for initial value problem and if $\phi(t)$ is uniformly stable, then the above condition is also satisfied.

Lemma 14.3. Assume that for any $(\psi,g) \in H(\phi,f)$, ψ is uni-formly stable with a common $\delta(\cdot)$. Then, for any $\varepsilon > 0$ and any $T > 0$ there exist positive numbers $\eta_1(\varepsilon)$ and $\eta_2(\varepsilon,T)$ such that for any $t_0 \geq 0$, if $|y(t_0)-\phi(t_0)| < \eta_1(\varepsilon)$ and $|h(t)| < \eta_2(\varepsilon,T)$ on $[t_0,t_0+T]$, then

$$|y(t)-\phi(t)| < \varepsilon \quad \text{on} \quad [t_0,t_0+T],$$

where $y(t)$ is a solution of (14.2) and $h(t)$ is continuous on I.

Proof. We can assume that $\varepsilon < B^*-B$. Let $K = \{x; |x| \leq \frac{B^*+B}{2}\}$. Then $|f(t,x)| \leq L$ on $I \times K$ for some constant $L > 0$. Letting $\eta_1(\varepsilon) = \frac{1}{2}\delta(\frac{\varepsilon}{2})$, suppose that for this $\eta_1(\varepsilon)$ there is no $\eta_2(\varepsilon,T)$. Then there exist $\{\tau_k\}$, $\tau_k \geq 0$, $\{t_k\}$, $\tau_k \leq t_k \leq \tau_k+T$, $\{h_k(t)\}$ and $\{y_k(t)\}$ such that

$$|\phi(\tau_k)-y_k(\tau_k)| < \frac{1}{2}\delta(\frac{\varepsilon}{2}), \quad |h_k(t)| < \frac{1}{k},$$

$$|\phi(t_k)-y_k(t_k)| = \varepsilon,$$

$$|\phi(t)-y_k(t)| < \varepsilon \quad \text{on} \quad [\tau_k,t_k),$$

where $y_k(t)$ is a solution on $[\tau_k,t_k]$ of $y' = f(t,y) + h_k(t)$. We can assume that $\{\phi(t+\tau_k),f(t+\tau_k,x)\}$ converges to $(\psi,g) \in H(\phi,f)$, $t_k-\tau_k \to \sigma$, $0 \leq \sigma \leq T$ and $y_k(t+\tau_k)$ converges to a continuous function $y(t)$ on $[0,\sigma]$. Then $y(t)$ is a solution of $x' = g(t,x)$ and $|y(0)-\psi(0)| \leq \frac{1}{2}\delta(\frac{\varepsilon}{2}) < \delta(\frac{\varepsilon}{2})$. But $|y(\sigma)-\psi(\sigma)| = \varepsilon$, which contradicts the uniform stability of $\psi(t)$.

Lemma 14.4. We assume that for every $(\psi,g) \in H(\phi,f)$, $\psi(t)$ is uniformly asymptotically stable with common triple $(\delta(\cdot),\delta_0,T(\cdot))$, that is, for any $\varepsilon > 0$, any $t_0 \geq 0$ and any $(\psi,g) \in H(\phi,f)$,

$$|\psi(t_0)-x_0| < \delta(\varepsilon) \quad \text{implies} \quad |\psi(t)-x(t)| < \varepsilon$$

$$\text{for all} \quad t \geq t_0$$

and

$$|\psi(t_0)-x_0| \leq \delta_0 \quad \text{implies} \quad |\psi(t)-x(t)| < \varepsilon$$

$$\text{for all} \quad t \geq t_0+T(\varepsilon),$$

whenever $x(t)$ is a solution of (14.10) through (t_0,x_0) . Then there
is an η_0 and for any $\varepsilon > 0$ and any $t_0 \geq 0$, there is an $\eta_3(\varepsilon) > 0$
such that if $|y(t_0)-\phi(t_0)| < \eta_0$ and if $|h(t)| < \eta_3(\varepsilon)$ on
$[t_0,t_0+T(\frac{\varepsilon}{2})]$, then the solution $y(t)$ of (14.2) is continuable on
$[t_0,t_0+T(\frac{\varepsilon}{2})]$ and

$$|y(t_0+T(\tfrac{\varepsilon}{2})) - \phi(t_0+T(\tfrac{\varepsilon}{2}))| < \varepsilon.$$

Proof. Since for any $(\psi,g) \in H(\phi,f)$, ψ is uniformly stable
with a common $\delta(\cdot)$, by Lemma 14.3, for any $\varepsilon > 0$ and any $T > 0$,
there exists an $\eta_1(\varepsilon) > 0$ and an $\eta_2(\varepsilon,T) > 0$. Suppose that $y(t_0)$
and $h(t)$ satisfy $|\phi(t_0)-y(t_0)| < \eta_1(\frac{B^*-B}{2})$ and $|h(t)| <$
$\eta_2(\frac{B^*-B}{2}, T(\frac{\varepsilon}{2}))$ on $[t_0,t_0+T(\frac{\varepsilon}{2})]$, then by Lemma 14.3, $y(t)$ is continuable to $t_0 + T(\frac{\varepsilon}{2})$. The other parts of the proof will be com-
pleted by the same argument as in the proof of Lemma 14.3. Namely,
letting $\eta_0 = \min(\delta_0,\eta_1(\frac{B^*-B}{2}))$, we can find a positive number
$\eta_3(\varepsilon) \leq \eta_2(\frac{B^*-B}{2},T(\frac{\varepsilon}{2}))$ such that for any $t_0 \geq 0$, $|\phi(t_0)-y(t_0)| < \eta_0$
and $|h(t)| < \eta_3(\varepsilon)$ on $[t_0,t_0+T(\frac{\varepsilon}{2})]$ imply $|y(t_0+T(\frac{\varepsilon}{2})) -$
$\phi(t_0+T(\frac{\varepsilon}{2}))| < \varepsilon$. This proves the lemma.

Lemma 14.5. In Lemmas 14.3 and 14.4, if $h(t)$ is locally
integrable on I and if

$$\int_{t_0}^{t_0+T} |h(s)|ds < \eta_2(\varepsilon,T) \quad \text{in Lemma 14.3}$$

and

$$\int_{t_0}^{t_0+T(\frac{\varepsilon}{2})} |h(s)|ds < \eta_3(\varepsilon) \quad \text{in Lemma 14.4,}$$

then the conclusions of the lemmas hold.

By using Lemmas 14.3 and 14.4, we shall prove the following theorem due to Kato [34].

Theorem 14.1.   For the almost periodic system (14.1), we assume that for every $(\psi,g) \in H(\phi,f)$, ψ is uniformly asymptotically stable with a common triple $(\delta(\cdot),\delta_0,T(\cdot))$.   Then the solution $\phi(t)$ is totally stable.   We can show also that $\phi(t)$ is totally asymptotically stable.   Here $\phi(t)$ is totally asymptotically stable if $\phi(t)$ is totally stable and if there exists a $\delta_0 > 0$ and for each $\varepsilon > 0$, there exists an $\eta(\varepsilon) > 0$ and a $T(\varepsilon) > 0$ such that if $|\phi(t_0)-y(t_0)| < \delta_0$, $t_0 \in I$, and $|h(t)| < \eta(\varepsilon)$ for $t \geq t_0$, then $|\phi(t)-y(t)| < \varepsilon$ for all $t \geq t_0+T(\varepsilon)$, where $y(t)$ is a solution of (14.2).

Proof.   Let η_1 and η_2 be the numbers given in Lemma 14.3 and let η_0 and η_3 be the numbers given in Lemma 14.4.   Let $\rho(\varepsilon) = \min\{\eta_1(\varepsilon),\eta_0\}$ and

$$\eta(\varepsilon) = \min\{\eta_2(\frac{B^*-B}{2},T(\frac{\rho(\varepsilon)}{2})),\ \eta_2(\varepsilon,T(\frac{\rho(\varepsilon)}{2})),\ \eta_3(\rho(\varepsilon))\}.$$

We shall prove that for any $t_0 \geq 0$, any solution $y(t)$ of (14.2) satisfies $|\phi(t)-y(t)| < \varepsilon$ for all $t \geq t_0 + T(\frac{\rho(\varepsilon)}{2})$, if $|\phi(t_0)-y(t_0)| < \eta_0$ and $|h(t)| < \eta(\varepsilon)$ on $[t_0,\infty)$.

Since we have

$$|\phi(t_0)-y(t_0)| < \eta_0 \leq \eta_1(\frac{B^*-B}{2}) \quad \text{and}$$

$$|h(t)| < \eta(\varepsilon) \leq \eta_2(\frac{B^*-B}{2},\ T(\frac{\rho(\varepsilon)}{2})),$$

it follows from Lemma 14.3 that

$$|\phi(t)-y(t)| < \frac{B^*-B}{2} \quad \text{on } t_0 \leq t \leq t_0 + T(\frac{\rho(\varepsilon)}{2}).$$

Moreover, by Lemma 14.4, we have

$$|\phi(t_0+T(\frac{\rho(\epsilon)}{2})) - y(t_0+T(\frac{\rho(\epsilon)}{2}))| < \rho(\epsilon), \tag{14.11}$$

because $|\phi(t_0)-y(t_0)| < \eta_0$ and $|h(t)| < \eta(\epsilon) \leq \eta_3(\rho(\epsilon))$. On the interval $t_0+T(\frac{\rho(\epsilon)}{2}) \leq t \leq t_0 + 2T(\frac{\rho(\epsilon)}{2})$, apply Lemma 14.3. Then, since we have (14.11) and $\rho(\epsilon) \leq \eta_1(\epsilon)$ and $|h(t)| < \eta(\epsilon) \leq \eta_2(\epsilon,T(\frac{\rho(\epsilon)}{2}))$, we have

$$|\phi(t)-y(t)| < \epsilon \quad \text{on} \quad t_0+T(\frac{\rho(\epsilon)}{2}) \leq t \leq t_0+2T(\frac{\rho(\epsilon)}{2}),$$

and moreover $\rho(\epsilon) \leq \eta_0$ and $\eta(\epsilon) \leq \eta_3(\rho(\epsilon))$ imply

$$|\phi(t_0+2T(\frac{\rho(\epsilon)}{2})) - y(t_0+2T(\frac{\rho(\epsilon)}{2}))| < \rho(\epsilon)$$

by Lemma 14.4.

By replacing $t_0 + T(\frac{\rho(\epsilon)}{2})$ by $t_0 + 2T(\frac{\rho(\epsilon)}{2})$ and by using the same argument, we can see that

$$|\phi(t)-y(t)| < \epsilon \quad \text{on} \quad t_0+2T(\frac{\rho(\epsilon)}{2}) \leq t \leq t_0 + 3T(\frac{\rho(\epsilon)}{2})$$

and

$$|\phi(t_0+3T(\frac{\rho(\epsilon)}{2})) - y(t_0+3T(\frac{\rho(\epsilon)}{2}))| < \rho(\epsilon).$$

Thus, by repeating the same argument, we have

$$|\phi(t)-y(t)| < \epsilon \quad \text{for all} \quad t \geq t_0 + T(\frac{\rho(\epsilon)}{2}).$$

If $|\phi(t_0)-y(t_0)| < \rho(\epsilon)$ and $|h(t)| < \eta(\epsilon)$, as will be seen from the above argument, we have $|\phi(t)-y(t)| < \epsilon$ on $t_0 \leq t \leq t_0 + T(\frac{\rho(\epsilon)}{2})$. Thus we can see the total stability. This completes the proof.

Corollary 14.2. For the almost periodic system (14.1), we assume that for each $g \in H(f)$, solutions are unique for initial value problem. If $\phi(t)$ is uniformly asymptotically stable, then it is totally stable.

This corollary follows immediately from Theorems 13.3 and 14.1. This corollary can be also proved by using Lemma 14.2 and similar arguments to those in the proof of Theorem 14.1 [36].

For a periodic system

$$x' = f(t,x), \quad f(t+\omega,x), \quad \omega > 0, \tag{14.12}$$

where $f \in C(R \times S_{B*}, R^n)$, we have the following result.

Corollary 14.3. For the periodic system, let $\phi(t)$ be a solution defined on I such that $|\phi(t)| \leq B$, $B < B^*$, for $t \geq 0$. If $\phi(t)$ is uniformly asymptotically stable, then it is totally stable.

This corollary follows immediately from Theorems 13.2 and 14.1.

The above results hold also for underlined functional differential equations and moreover, a more general system. Namely, let $f(t,x)$ be a continuous function defined on $[0,\infty) \times S_{B*}$ and let $T(f)$ be the set of continuous functions $f(t+\tau,x)$ for all $\tau \geq 0$, which is a subset of $C(I \times S_{B*}, R^n)$, where $C(I \times S_{B*}, R^n)$ denotes the set of all continuous R^n-valued functions defined on $I \times S_{B*}$ and it is a topological space by the compact-open topology. Let $H(f)$ be the closure of $T(f)$. Then, if we assume that $f(t,x)$ is uniformly continuous in (t,x) and bounded on $I \times S$ for any compact set S in S_{B*}, $H(\phi)$ and $H(\phi,f)$ are compact, where $\phi(t)$ is a solution such that $|\phi(t)| \leq B$, $B < B^*$, for $t \geq 0$. Therefore, under this assumption, we can prove the above results. For the details, see [34], [36]. Theorem 13.4 also holds for this family of functions, but for ordinary differential equations [52].

The following results were informed by Shui-Nee Chow. Namely, by using Lemma 14.5 and by the same argument, we have the following results.

Theorem 14.2. For the almost periodic system (14.1), we as-
sume that for every $(\psi, g) \varepsilon H(\phi, f)$, ψ is uniformly asymptotically
stable with a common triple $(\delta(\cdot), \delta_0, T(\cdot))$. Then the solution $\phi(t)$
is integrally attracting and consequently it is integrally asymptoti-
cally stable.

Corollary 14.4. For the almost periodic system (14.1), we assume that for each $g \varepsilon H(f)$, solutions are unique for initial value
problem. If $\phi(t)$ is uniformly asymptotically stable, then it is
integrally asymptotically stable. Notice that if $\phi(t)$ is integrally
asymptotically stable, it is uniformly asymptotically stable.

Remark. The zero solution of (13.9) is not integrally asymptotically stable.

Corollary 14.5. For the periodic system (14.12), the uniformly
asymptotic stability of $\phi(t)$ is equivalent to the integrally
asymptotic stability.

Theorem 14.3. For the periodic system (14.12), if $\phi(t)$ is
uniformly asymptotically stable, there exists a Liapunov function
$V(t,x)$ defined on $0 \leq t < \infty$, $|\phi(t)-x| < \alpha$ for some α, which satisfies the following conditions;

> (i) $a(|\phi(t)-x|) \leq V(t,x) \leq |\phi(t)-x|$, where $a(r)$ is con-
>   tinuous and positive definite,
> (ii) $|V(t,x)-V(t,y)| \leq |x-y|$,
> (iii) $\dot{V}_{(14.12)}(t,x) \leq -V(t,x)$.

Here it should be noticed that we assume only the continuity
of $f(t,x)$.

CHAPTER III

EXISTENCE THEOREMS FOR PERIODIC SOLUTIONS

AND ALMOST PERIODIC SOLUTIONS

15. Existence Theorems for Periodic Solutions

First of all, we shall state some fixed point theorems without proofs. The following theorem is due to <u>Brouwer</u>. For the proof, see [5].

Theorem 15.1. Let G be a simply connected plane open domain and T a topological mapping of G into itself. If T is sense-preserving and there exists a point $x_0 \varepsilon G$ and a subsequence of the successive images $x_1 = Tx_0$, $x_2 = Tx_1, \ldots$ which converges to a point in G , then T has a fixed point in G .

The following theorem due to <u>Browder</u> is useful in obtaining existence theorems for periodic solutions. For the proof, see [6].

Theorem 15.2. Let S and S_1 be open convex subsets of the Banach space X and let S_0 be a closed convex subset of X such that $S_0 \subset S_1 \subset S$. If T is a continuous mapping of S into X such that $T(S)$ is contained in a compact set of X and if for a positive integer m , T^m is well-defined on S_1 and $\underset{0 \le j \le m}{\cup} T^j(S_0) \subset S_1$ while $T^m(S_1) \subset S_0$, then T has a fixed point in S_0 .

Now consider a periodic system

$$x' = f(t,x), \quad f(t+\omega,x) = f(t,x), \quad \omega > 0, \qquad (15.1)$$

where $f(t,x) \varepsilon C(R \times R^n, R^n)$ and ω is the period of $f(t,x)$. Since we shall use some fixed point theorems, we assume the unique-

ness of solutions of (15.1) throughout this section. Massera proved
the following theorems [46].

Theorem 15.3. If (15.1) is a scalar equation, the existence of
a solution which exists and remains bounded in the future implies the
existence of a periodic solution of period ω.

Proof. Let $\phi(t)$ be the bounded solution of (15.1), where
we can assume that $\phi(t)$ is defined on I. Let $\phi_k(t) = \phi(t+k\omega)$,
where $k > 0$ is an integer. Then $\phi_k(t)$ is also a solution of (15.1).
If $\phi(0) = \phi_1(0)$, $\phi(t)$ is a periodic solution of period ω. Now assume that $\phi(0) < \phi_1(0)$. In this case, we can see that $\phi_k(0) <$
$\phi_{k+1}(0)$ for all k, because of the uniqueness. Similarly, if
$\phi(0) > \phi_1(0)$, then $\phi_k(0) > \phi_{k+1}(0)$ for all k. Therefore the sequence $\{\phi_k(t)\}$ is monotone, and clearly it is uniformly bounded
and equicontinuous on $[0,\omega]$. This implies that $\{\phi_k(t)\}$ converges
to a function $\psi(t)$ uniformly on $[0,\omega]$. Clearly $\psi(t)$ is a solution
of (15.1) defined on $[0,\omega]$ and $\phi_k(0) \to \psi(0)$, $\phi_k(\omega) = \phi_{k+1}(0) \to \psi(\omega)$.
Thus $\psi(0) = \psi(\omega)$, and hence we can see the existence of a periodic
solution of period ω.

Theorem 15.4. Consider a linear system

$$x' = A(t)x + b(t), \qquad\qquad (15.2)$$

where A(t) is an $n \times n$ continuous matrix defined on R and b(t)
is a continuous n-vector defined on R and $A(t+\omega) = A(t)$,
$b(t+\omega) = b(t)$. Then the existence of a solution which is bounded in
the future implies the existence of a periodic solution of period ω.

Proof. If $b(t) \equiv 0$, the existence of a periodic solution is
clear. Assume that $b(t) \not\equiv 0$. Let x_0 be any vector and let
$x_1 = x(\omega,0,x_0)$, where $x(t,0,x_0)$ is the solution of (15.2) through

$(0,x_0)$. We have

$$x_1 = x(\omega,0,x_0) = X(\omega)X^{-1}(0)x_0 + \int_0^\omega X(\omega)X^{-1}(s)b(s)ds,$$

where $X(t)$ is a fundamental matrix of $x' = A(t)x$.   Setting
$P = X(\omega)X^{-1}(0)$ and $q = \int_0^\omega X(\omega)X^{-1}(s)b(s)ds$, consider the transformation T such that

$$x_1 = Tx_0 = Px_0 + q.$$

Suppose that the system of linear equations

$$(P-E)x + q = 0 \tag{15.3}$$

has no solution, where E is the unit matrix.   Then $P-E$ is singular
and there exists a fixed vector y such that $y'(P-E) = 0$ and
$y'q \neq 0$, where y' is the transpose of y.

Since $y'P = y'$, we have $y'P^k = y'$.   Therefore, if we apply
successively the transformation T, we have

$$x_k = T^k x_0 = P^k x_0 + (P^{k-1}+P^{k-2}+\ldots+E)q,$$

and hence $y'x_k = y'x_0+ky'q$.   Since $y'q \neq 0$, we have $y'x_k \to \infty$ as
$k \to \infty$.   If x_0 is the initial point of the bounded solution,
$x_k = x(k\omega,0,x_0)$ and hence $y'x_k$ remains bounded.   Thus there arises
a contradiction.   Therefore the linear equation (15.3) has a solution
x^*, that is, $(P-E)x^* + q = 0$ or $x^* = Px^* + q$, which means
$x^* = x(\omega,0,x^*)$.   Thus we see the existence of a periodic solution of
period ω.

For _functional differential equations_, Theorem 15.3 does not
hold, see [25].   But Theorem 15.4 also holds, see [9], [10].

Massera [46] gave an example which shows that if $n = 2$ in
the system (15.1), the existence of a bounded solution does not

necessarily imply the existence of a <u>periodic solution of period</u> ω.

Example 15.1. Consider a system

$$\begin{cases} x' = f(u,v)\cos^2 \pi t - g(u,v)\sin \pi t \cos \pi t - \pi y \\ y' = g(u,v)\cos^2 \pi t + f(u,v)\sin \pi t \cos \pi t + \pi x, \end{cases} \qquad (15.4)$$

where $x \in R$, $y \in R$ and

$$\begin{cases} u = x \cos \pi t + y \sin \pi t \\ v = y \cos \pi t - x \sin \pi t. \end{cases} \qquad (15.5)$$

The following assumptions will be made:

(a) f and g have continuous first partial derivatives.

(b) $f(-u,-v) = f(u,v)$ and $g(-u,-v) = g(u,v)$.

(c) $f(1,0) = g(1,0) = 0$, $f(0,v) = 0$, $g(0,v) > 0$ for all v.

(d) $\int_{-\infty}^{\infty} \dfrac{1}{g(0,v)}\, dv < \dfrac{2}{\pi}$.

For example, $f = uv$, $g = (1-u^2)(1+v^2)c$ satisfy the conditions, where c is a suitable constant, for instance, $\dfrac{\pi^2}{2} < c$.

Then condition (a) implies the uniqueness of solutions and we can easily see that system (15.4) is periodic of period 1, because of assumption (b). In the variables (u,v), system (15.4) becomes

$$u' = f(u,v)\cos \pi t, \quad v' = g(u,v)\cos \pi t. \qquad (15.6)$$

Since $u = \pm 1$, $v = 0$ are solutions of (15.6), $\{x = \cos \pi t, y = \sin \pi t\}$ and $\{x = -\cos \pi t, y = -\sin \pi t\}$ are two periodic solutions of (15.4) of period 2. Clearly $u(t) \equiv 0$, $v = v(t)$ is a solution of (15.6) and hence, if $u > 0$ at any point t_0 of any solution of (15.6), we have $u > 0$ for all t, because of uniqueness. These solutions cannot be a solution $\{x(t),y(t)\}$ of period 1, because if $\{x(t),y(t)\}$ is a solution of period 1, $u(t+1) = x(t+1)\cos(\pi t + \pi) +$

$y(t+1)\sin(\pi t+\pi) = -x(t)\cos \pi t - y(t)\sin \pi t$, which shows that u must change signs. The same is true if $u < 0$ at any point t_0. Finally, suppose that $u = 0$, $v = v(t)$ is a certain solution of (15.6). The $v(t)$ is given by

$$\int_{v_0}^{v} \frac{dv}{g(0,v)} = \int_{t_0}^{t} \cos \pi t\, dt = \frac{1}{\pi}(\sin \pi t - \sin \pi t_0).$$

When t increases from $-1/2$ to $1/2$, we have

$$\int_{v(-1/2)}^{v(1/2)} \frac{dv}{g(0,v)} = \frac{2}{\pi},$$

which contradicts (d). Hence $v(t)$ cannot be defined for all t and the corresponding solution $\{x(t),y(t)\}$ cannot be periodic (cannot exist in the future).

For the case $n = 2$, we have the following theorem [46].

Theorem 15.5. If $n = 2$ in the system (15.1) and if all solutions exist in the future and one of them is bounded, then there exists a periodic solution of period ω.

Proof. Let T be a mapping such that $x_0 \rightarrow x(\omega,0,x_0)$. Since we assume the uniqueness of solutions and all solutions starting at $t = 0$ can reach $t = \omega$, T is a topological mapping of a plane into itself and T is sense-preserving. If x_0 is the starting point of the bounded solution $\phi(t)$, its successive images are $\phi(\omega)$, $\phi(2\omega),\ldots,$ which form a bounded sequence and consequently have a convergent subsequence. Applying Theorem 15.1 to G, where G is the whole plane, T has a fixed point $x^* \in G$, that is $x^* = x(\omega,0,x^*)$. Thus we can see that $x(t,0,x^*)$ is a periodic solution of period ω.

Corollary 15.1. If $n = 2$ in system (15.1) and if all solutions of (15.1) are bounded, then there exists a periodic solution of period ω.

Example 15.2. Consider the equation

$$x'' + f(x)x' + g(x) = p(t) \qquad\qquad (15.7)$$

and assume that

(a) $f(x)$ is continuous on R and $F(x) = \int_0^x f(u)\,du \to \pm\infty$
as $x \to \pm\infty$, respectively,

(b) $g(x)$ satisfies locally a Lipschitz condition, and
$xg(x) \geq 0$ for $|x| > q$,

(c) $p(t)$ is continuous on R and is periodic in t of
period ω, and $\int_0^\omega p(t)\,dt = 0$.

Then equation (15.7) has a periodic solution of period ω.

To see this, consider an equivalent system

$$x' = y - F(x) + P(t), \quad y' = -g(x), \qquad (15.8)$$

where $P(t) = \int_0^t p(s)\,ds$. By the condition $\int_0^\omega p(t)\,dt = 0$, $P(t)$ is
also periodic of period ω. Therefore, as was seen in Example 8.5,
every solution of (15.8) is bounded, and hence, by Corollary 15.1,
system (15.8) has a periodic solution of period ω. Thus equation
(15.7) has a periodic solution of period ω.

Remark. In case $n > 2$, Theorem 15.5 is not necessarily true
(cf. [46]). Another example which shows this fact will be found
later in this section.

Now we shall discuss the existence of a bounded solution of

$$x'' = f(t,x,x'), \qquad\qquad (15.9)$$

where $f(t,x,y) \in C(R \times R \times R, R)$ and $f(t,x,y)$ is periodic in t
of period ω. If an equivalent system

$$x' = y, \quad y' = f(t,x,y) \qquad\qquad (15.10)$$

has a bounded solution and if every solution is continuable to any

$T \varepsilon (0, \infty)$ (only to the right), we can apply Theorem 15.5 to show

the existence of a periodic solution. Concerning the continuation

of solution, the following theorem is more useful in some cases. For

the proof, see [74]. Also, for global existence, see [35].

Theorem 15.6. Consider a system

$$x' = f(t,x,y), \quad y' = g(t,x,y), \qquad (15.11)$$

where $f \varepsilon C(I \times R^n \times R^m, R^n)$ and $g \varepsilon C(I \times R^n \times R^m, R^m)$. Suppose

that for each $T \varepsilon (0, \infty)$, there exists a Liapunov function $W(t,x,y)$

defined on $0 \leq t \leq T$, $|x| + |y| \geq K$, where K can be large, which

satisfies the following conditions;

    (i) $W(t,x,y) \to \infty$ uniformly for t,x as $|y| \to \infty$,

    (ii) $\dot{W}_{(15.11)}(t,x,y) \leq 0$.

Moreover, suppose that for each L and each T, there exists a

Liapunov function $U(t,x,y)$ defined on $0 \leq t \leq T$, $|x| \geq K^*$, $|y| \leq L$,

where K^* can be large, which satisfies the following conditions;

    (iii) $U(t,x,y) \to \infty$ uniformly for t,y as $|x| \to \infty$,

    (iv) $\dot{U}_{(15.11)}(t,x,y) \leq 0$.

Then all solutions of (15.11) exist in the future.

Theorem 15.7. Consider the periodic equation (15.9). Suppose

that there exist two functions $\alpha(t)$ and $\beta(t)$ defined on I, twice

differentiable and bounded on I with their derivatives and suppose

that $\alpha(t) \leq \beta(t)$, $\alpha''(t) \geq f(t,\alpha(t),\alpha'(t))$ and $\beta''(t) \leq$

$f(t,\beta(t),\beta'(t))$. Let D be the domain $0 \leq t < \infty$, $\alpha(t) \leq x \leq \beta(t)$,

and let $D_1 = \{(t,x,y); (t,x) \varepsilon D, y \geq K\}$ and

$D_2 = \{(t,x,y); (t,x) \varepsilon D, y \leq -K\}$, where $K > 0$ can be large.

Moreover, suppose that there are two Liapunov functions $V_i(t,x,y)$ defined on D_i, $i = 1,2$, respectively, which satisfy the following conditions;

(i) $V_i(t,x,y) \le b(|y|)$, $i = 1,2$, where $b(r) > 0$ is continuous,

(ii) $V_i(t,x,y) \to \infty$ uniformly for (t,x) as $|y| \to \infty$,

(iii)   in the interiors of D_1 and D_2,

$$\dot{V}_1(t,x,y) = \overline{\lim_{h\to 0^+}} \frac{1}{h}\{V_1(t+h,x+hy,y+hf(t,x,y))-V_1(t,x,y)\} \ge 0,$$

$$\dot{V}_2(t,x,y) = \overline{\lim_{h\to 0^+}} \frac{1}{h}\{V_2(t+h,x+hy,y+hf(t,x,y))-V_2(t,x,y)\} \le 0.$$

Then the equation (15.9) has a solution $x(t)$ such that $|x(t)| + |x'(t)|$ is bounded for all $t \ge 0$.

Proof. Let n be a positive integer and consider the two point boundary value problem on each interval $[0,n]$. Since all conditions in Corollary 5.1 are satisfied, we can see that there exists a solution $x_n(t)$ of (15.9) which satisfies the conditions

$$x_n(0) = \alpha(0) \quad \text{and} \quad x_n(n) = \alpha(n),$$

and $\alpha(t) \le x_n(t) \le \beta(t)$, $|x_n'(t)| \le M$ on $[0,n]$ for all $n \ge 1$. Here, as is seen from the proof of Corollary 5.1, M can be chosen so that M is independent of n and $|\alpha'(t)| < M$. Let $y_n(t)$ be

$$y_n(t) = \begin{cases} x_n(t) & (0 \le t \le n) \\ \alpha(t) & (n < t < \infty). \end{cases}$$

Then the sequence $\{y_n(t)\}$ is uniformly bounded and equicontinuous, and hence a subsequence converges to some function $x(t)$ uniformly on any compact interval on I. Clearly $\alpha(t) \le x(t) \le \beta(t)$ on I and $|x'(t)| \le M$, and hence $x(t)$ is a bounded solution of (15.9).

Example 15.3. Consider the equation

$$x'' + f(x,x') + g(t,x) = p(t), \qquad (15.12)$$

where we assume that

(a) $f(x,y)$ satisfies locally a Lipschitz condition, $g(t,x)$ is locally Lipschitzian in x and $p(t)$ is continuous on R,

(b) $g(t,x)$ and $p(t)$ are periodic in t of period ω,

(c) $f(x,y)y \geq 0$,

(d) $\int_0^x g(t,u)\,du = G(t,x) > -c$ for all t,x, where $c > 0$ is a constant, and $|\frac{\partial G}{\partial t}|/\sqrt{G(t,x)+c}$ is bounded,

(e) there exist $a,b, a < b$, such that

$$\begin{cases} 0 \leq f(a,0) + g(t,a) - p(t) \\ 0 \geq f(b,0) + g(t,b) - p(t). \end{cases} \qquad (15.13)$$

Furthermore, we assume that there exists a continuous function $\lambda(u) > 0$ for $-\infty < u < \infty$ such that $|f(x,y)| \leq \lambda(y)$, $\int^{-\infty} \frac{u}{\lambda(u)+m}\,du = \infty$ and $\int^{\infty} \frac{u}{\lambda(u)+m}\,du = \infty$, where $|g(t,x)|+|p(t)| \leq m$ for $t \in R$ and $x \in [a,b]$. Then the equation (15.12) has a periodic solution of period ω.

To see this, for an arbitrary $T > 0$, consider a system

$$x' = y, \quad y' = -f(x,y)-g(t,x) + p(t) \qquad (15.14)$$

on the domain $0 \leq t \leq T$, $|x| < \infty$, $|y| < \infty$. On the domain $0 \leq t \leq T$, $x^2 + y^2 \geq K^2$, consider a function

$$W(t,x,y) = \exp\{ \sqrt{2(G(t,x)+c)+y^2} - \int_0^t |p(s)|\,ds-kt\},$$

where $|\frac{\partial G}{\partial t}|/\sqrt{2(G(t,x)+c)} \leq k$. Then we have $\dot{W}_{(15.14)}(t,x,y) \leq 0$.

By using W(t,x,y), we can show the boundedness of $|y(t)|$, which
implies the boundedness of $|x(t)|$. Thus it can be seen that all
solutions of (15.14) are continuable to t = T. Since T is arbitrary, all solutions exist in the future.

Letting $\alpha(t) = a$ and $\beta(t) = b$, if $V_1(t,x,y)$ and $V_2(t,x,y)$
are defined by

$$V_1(t,x,y) = \exp\{x + \int_K^y \frac{u}{\lambda(u)+m}\, du\},$$

$$V_2(t,x,y) = \exp\{x + \int_{-K}^y \frac{u}{\lambda(u)+m}\, du\},$$

all conditions in Theorem 15.7 are satisfied. Thus, applying
Theorem 15.5, we see the existence of a periodic solution of (15.12).

In a special equation $x" + k \sin x = p(t)$, k > 0, if p(t)
is periodic of period ω and $|p(t)| \leq k$, then the equation has a
periodic solution of period ω.

Now we shall consider the case where n in the system (15.1)
is an arbitrary positive integer. By applying Theorem 15.2, we obtain
the following theorem, which corresponds to Cartwright's theorem for
second order equations [8].

Theorem 15.8. If the solutions of (15.1) are ultimately
bounded for bound B, then there exists a periodic solution x(t) of
period ω such that $|x(t)| \leq B$ for all t.

Proof. Since we assume the uniqueness of solutions, by
Theorem 8.5, the solutions of (15.1) are uniformly bounded and uniformly ultimately bounded for bound H for some H $\geq$ B. Let T be
a mapping such that

$$Tx_0 = x(\omega,0,x_0).$$

Since the solutions are uniformly bounded, there exists a $\beta(H) > 0$ such that if $t_0 \varepsilon I$ and $x_0 \varepsilon \overline{S}_H$, $S_H = \{x; |x| < H\}$, then $|x(t,t_0,x_0)| < \beta$ for all $t \geq t_0$. Moreover, there are γ, γ^* such that $t_0 \varepsilon I$ and $x_0 \varepsilon \overline{S}_\beta$ imply $|x(t,t_0,x_0)| < \gamma$ for all $t \geq t_0$ and that $t_0 \varepsilon I$ and $x_0 \varepsilon \overline{S}_\gamma$ imply $|x(t,t_0,x_0)| < \gamma^*$ for all $t \geq t_0$. Let $S = S_\gamma$. Then $T(S)$ is contained in a compact set $\overline{S}_{\gamma^*}$, and it is clear that T is continuous. From uniformly ultimate boundedness for bound H, it follows that there exists a $\tau > 0$ such that if $t \geq \tau$ and $|x_0| < \beta$, then $|x(t,0,x_0)| < H$, and hence there is a positive integer m for which $|x(m\omega,0,x_0)| < H$ if $|x_0| < \beta$.

Let $\overline{S}_H$ be S_0 in Theorem 15.2 and S_β be S_1 in Theorem 15.2. Then these convex sets satisfy the assumptions in Theorem 15.2. Therefore there exists a fixed point x_0 in $\overline{S}_H$, which implies the existence of a periodic solution $x(t)$ of period ω. Since $x(t)$ is bounded by B for all large t and is periodic, clearly $|x(t)| \leq B$ for all $t \varepsilon R$. This completes the proof.

The most general result for periodic processes has been obtained by Hale, LaSalle and Slemrod by assuming that the system is dissipative [30]. Also, see [31].

Example 15.4. Consider the equation of third order

$$x''' + \phi(x')x'' + bx' + f(x) = p(t), \qquad (15.15)$$

where $b > 0$ is a constant, $\phi(y)$ is continuous, $f(x)$ is locally Lipschitzian, $p(t)$ is continuous and periodic of period ω and $\int_0^\omega p(s)ds = 0$. Under the following assumptions;

  (i) $|f(x)| \leq F$ for all x, and $xf(x) > 0$ for $|x| \geq h$,

  (ii) $y\Phi(y) > 0$ for $|y| \geq k$, and $|\Phi(y)| \to \infty$ as $|y| \to \infty$,
      where $\Phi(y) = \int_0^y \phi(u)du$,

we have seen in Section 8 that the solutions of an equivalent system

$$x' = y, \quad y' = z-\phi(y) + P(t), \quad z' =-f(x) - by \qquad (15.16)$$

are uniformly ultimately bounded. Therefore, by Theorem 15.8, the
system (15.16) and consequently the equation (15.15) has a periodic
solution of period ω.

Now we shall discuss a perturbed system with a small periodic
perturbation

$$x' = f(t,x) + g(t,x,\varepsilon). \qquad (15.17)$$

Suppose that system (15.1) has a periodic solution $\phi(t)$ of period ω
and $f(t,x)$ is defined and continuous on D: $-\infty < t < \infty$, $|x-\phi(t)| < A$
and $f(t,x)$ is locally Lipschitzian in x. Furthermore, suppose that
$g(t,x,\varepsilon)$ is continuous in (t,x,ε) for $(t,x) \in D$ and $|\varepsilon| \le \varepsilon^*$,
is locally Lipschitzian in x, is periodic in t of period ω and
satisfies the condition $|g(t,x,\varepsilon)| \le \lambda(t,\varepsilon)$ for $(t,x) \in D$, where
$\lambda(t,\varepsilon) = 0(|\varepsilon|)$ uniformly in t, $t \ge 0$, as $|\varepsilon| \to 0$.

Lemma 15.1. Under the assumptions above, if $\phi(t)$ is uniformly asymptotically stable, for each η, $0 < \eta < A$, there exists a
$\delta(\eta)$ and an $\varepsilon_0(\eta)$ such that $|x_0-\phi(t_0)| \le \delta(\eta)$ and $|\varepsilon| \le \varepsilon_0(\eta)$
imply $|x(t,t_0,x_0,\varepsilon)-\phi(t)| < \eta$ for all $t \ge t_0$, where $x(t,t_0,x_0,\varepsilon)$
is a solution of (15.17) through (t_0,x_0).

Proof. Since $\lambda(t,\varepsilon) = 0(|\varepsilon|)$ as $|\varepsilon| \to 0$, if $|\varepsilon|$ is small,
then $|g(t,x,\varepsilon)|$ is small. Moreover, by Corollary 14.3, $\phi(t)$ is
totally stable, and hence the lemma follows immediately from the
definition of total stability.

Theorem 15.9. Under the assumptions above, if $\phi(t)$ is uni-
formly asymptotically stable and $|\varepsilon|$ is sufficiently small, system

(15.17) has a periodic solution $\phi(t,\varepsilon)$ of period ω, and we can select such a solution $\phi(t,\varepsilon)$ which tends to $\phi(t)$ as $\varepsilon \to 0$.

Proof. Since $\phi(t)$ is uniformly asymptotically stable, by Theorem 14.3, for some $\alpha < A$ there exists a Liapunov function $V(t,x)$ defined on $0 \le t < \infty$, $|\phi(t)-x| < \alpha$, which satisfies

(i) $a(|\phi(t)-x|) \le V(t,x) \le |\phi(t)-x|$, where $a(r)$ is continuous, increasing and positive definite,

(ii) $|V(t,x)-V(t,y)| \le |x-y|$,

(iii) $\dot{V}_{(15.1)}(t,x) \le -V(t,x)$.

For any small $\eta > 0$, let $\delta(\eta)$ and $\varepsilon_0(\eta)$ be the numbers in Lemma 15.1. Here we can assume that ε_0 is so small that $\lambda(t,\varepsilon) < \frac{1}{2}a(\delta(\eta))$, if $|\varepsilon| \le \varepsilon_0(\eta)$. Thus, if $|\varepsilon| \le \varepsilon_0(\eta)$, we have

$$\dot{V}_{(15.17)}(t,x) \le -V(t,x) + \lambda(t,\varepsilon)$$
$$\le -a(\delta(\eta)) + \frac{1}{2}a(\delta(\eta)) = -\frac{1}{2}a(\delta(\eta))$$

on the domain $0 \le t < \infty$, $|\phi(t)-x| \ge \delta(\eta)$. Therefore we can see that there exists a $T(\eta) > 0$ such that if $|\phi(0)-x_0| < \eta$, then $|x(t,0,x_0,\varepsilon)-\phi(t)| \le \delta(\eta)$ for all $t \ge T(\eta)$. Thus there is a positive integer m such that $|x(m\omega,0,x_0,\varepsilon)-\phi(m\omega)| \le \delta(\eta)$. Furthermore, by Lemma 15.1, $|x(t,t_0,x_0,\varepsilon)-\phi(t)| < \eta$ for all $t \ge t_0$ if $|\phi(0)-x_0| \le \delta(\eta)$ and $|\varepsilon| \le \varepsilon_0(\eta)$. Thus, by choosing η so that $\eta < \delta(\alpha)$ and applying Theorem 15.2, it is shown that system (15.17) has a periodic solution $\phi(t,\varepsilon)$ of period ω if $|\varepsilon| \le \varepsilon_0(\eta)$. Moreover, it is clear that a periodic solution $\phi(t,\varepsilon)$ of (15.17) stays in the domain $-\infty < t < \infty$, $|\phi(t)-x| \le \eta$. Therefore we obtain a periodic solution of (15.17) as close to $\phi(t)$ as desired, if $|\varepsilon|$ is small. This completes the proof.

For more general results, see [23], [33], [77], [78].

Now we consider a periodic system which has a bounded solution with some stability property.

Theorem 15.10.   Suppose that $f(t,x)$ in (15.1) is continuous on $R \times S_{B*}$, $S_{B*} = \{x; |x| < B*\}$ and the periodic system (15.1) has a solution $\phi(t)$ such that $|\phi(t)| \leq B$, $B < B*$, for all $t \geq 0$.   If there is a $\delta > 0$, such that $|\phi(t_0)-x_0| < \delta$ implies $|\phi(t)-x(t,t_0,x_0)| \to 0$ as $t \to \infty$, then there exists a periodic solution of (15.1) of period $m\omega$, where $m \geq 1$ is some integer and $x(t,t_0,x_0)$ is a solution of (15.1) through (t_0,x_0).   Note that we do not assume the uniqueness of solutions.

Proof.   Let $\phi_k(t) = \phi(t+k\omega)$, where k is a positive integer. Then $\phi_k(t)$ is a solution of (15.1) through $(0,\phi(k\omega))$ and $|\phi_k(t)| \leq B$ for $t \geq 0$.   Since $\{\phi_k(t)\}$ is uniformly bounded and equicontinuous, there is a subsequence $\{\phi_{k_j}(t)\}$ such that $|\phi_{k_1}(0)-\phi_{k_2}(0)| < \delta$, $k_2 > k_1$, and that $\phi_{k_j}(t)$ converges to a solution $\psi(t)$ of (15.1) uniformly on any compact set in I.

Set $m = k_2-k_1$ and consider a solution $\phi(t+m\omega)$.   We have

$$|\phi(k_1\omega+m\omega)-\phi(k_1\omega)| = |\phi(k_2\omega)-\phi(k_1\omega)| = |\phi_{k_2}(0)-\phi_{k_1}(0)| < \delta.$$

Therefore, by the assumption, we have

$$|\phi(t)-\phi(t+m\omega)| \to 0 \quad \text{as} \quad t \to \infty. \qquad (15.18)$$

Since $\phi_{k_j}(t) \to \psi(t)$ as $j \to \infty$, we have

$$\phi(m\omega+k_j\omega) = \phi_{k_j}(m\omega) \to \psi(m\omega) \quad \text{as} \quad j \to \infty$$

or

$$\phi_{k_j+m}(0) \to \psi(m\omega) \quad \text{as} \quad j \to \infty. \qquad (15.19)$$

On the other hand, by (15.18),

$$|\phi(k_j\omega+m\omega)-\phi(k_j\omega)| \to 0 \quad \text{as} \quad j \to \infty$$

or

$$|\phi_{k_j+m}(0)-\phi_{k_j}(0)| \to 0 \quad \text{as} \quad j \to \infty, \qquad (15.20)$$

because $k_j\omega \to \infty$. Since $\phi_{k_j}(0) \to \psi(0)$ as $j \to \infty$ and we have (15.20),

$$|\phi_{k_j+m}(0)-\phi(0)| \to 0 \quad \text{as} \quad j \to \infty.$$

From this and (15.19), it follows that $\psi(m\omega) = \psi(0)$. Thus we can construct a periodic solution of (15.1) of period $m\omega$.

Remark. If every solution of (15.1) which remains in $\overline{S}_B$ in the future tends to $\phi(t)$ as $t \to \infty$, then clearly we have $|\phi(t+\omega)-\phi(t)| \to 0$ as $t \to \infty$. Therefore we have $\psi(\omega) = \psi(0)$ and consequently there exists a periodic solution of period ω.

The following example due to Chow shows that the existence of a bounded uniformly asymptotically stable solution does not necessarily imply the existence of a periodic solution of period ω. Also every solution of this example is bounded.

Let A_{-1}, A_0, A_1 be the sets in the $(x-y)$-plane and let B_1, B_2 be sets in (x,y,z)-space. They are given by

$$A_{-1} = \{(x,y); (x+4)^2 + y^2 \leq 1\},$$
$$A_0 = \{(x,y); x^2 + y^2 \leq 1\},$$
$$A_1 = \{(x,y); (x-4)^2 + y^2 \leq 1\},$$
$$B_1 = \{(x,y,z); x^2 + y^2 = 1, |z-5k| \leq 1 \text{ for some integer } k\},$$
$$B_2 = \{(x,y,z); x^2 + y^2 \geq 1, z = 5k-1 \text{ or } 5k \text{ or } 5k+1 \text{ for}$$
$$\text{some integer } k\}.$$

Letting $A = A_{-1} \cup A_0 \cup A_1$ and $B = B_1 \cup B_2$, we define functions $\alpha(x,y)$ and $\beta(x,y,z)$ by

$$\alpha(x,y) = \inf\{\sqrt{(x-u)^2 + (y-v)^2}; \quad (u,v) \; \varepsilon \; R^2 - A\},$$

$$\beta(x,y,z) = \frac{\rho(x,y,z)}{1+\rho(x,y,z)},$$

where $\rho(x,y,z) = \inf\{\sqrt{(x-u)^2 + (y-v)^2 + (z-w)^2}; \quad (u,v,w) \; \varepsilon \; B\}$. Define functions $\phi(x,y)$ and $\psi(x,y)$ by

$$\phi(x,y) = \begin{cases} [-(x+4)+y]\alpha(x,y) & \text{if} \quad (x,y) \; \varepsilon \; A_{-1} \\ [-x+y]\alpha(x,y) & \text{if} \quad (x,y) \; \varepsilon \; A_0 \\ [-(x-4)+y]\alpha(x,y) & \text{if} \quad (x,y) \; \varepsilon \; A_1 \\ 0 & \text{if} \quad (x,y) \; \overline{\varepsilon} \; A \end{cases}$$

and

$$\psi(x,y) = \begin{cases} -(x+4)\alpha(x,y) & \text{if} \quad (x,y) \; \varepsilon \; A_{-1} \\ -x\alpha(x,y) & \text{if} \quad (x,y) \; \varepsilon \; A_0 \\ -(x-4)\alpha(x,y) & \text{if} \quad (x,y) \; \varepsilon \; A_1 \\ 0 & \text{if} \quad (x,y) \; \overline{\varepsilon} \; A \end{cases}$$

Then for the system $x' = \phi(x,y)$, $y' = \psi(x,y)$, the points $(-4,0)$, $(0,0)$, $(4,0)$ are spiral points and all points in the closure of the complement of A are critical points. Now we construct a system in (x,y,z)-space,

$$x' = f^*(x,y,z)$$
$$y' = g^*(x,y,z)$$
$$z' = h^*(x,y,z),$$

where

$$f^*(x,y,z) = \begin{cases} \phi(x,y)[2-(z-5k)] & \text{on} \quad 0 \leq z-5k \leq 2 \\ 16^2[3-(z-5k)][(z-5k)-2] & \text{on} \quad 2 \leq z-5k \leq 3 \\ \phi(x,y)[(z-5k)-3] & \text{on} \quad 3 \leq z-5k \leq 5, \end{cases}$$

$$g^*(x,y,z) = \begin{cases} \psi(x,y)[2-(z-5k)] & \text{on } 0 \leq z-5k \leq 2 \\ 0 & \text{on } 2 \leq z-5k \leq 3 \\ \psi(x,y)[(z-5k)-3] & \text{on } 3 \leq z-5k \leq 5 \end{cases}$$

and

$$h^*(x,y,z) = \begin{cases} -\beta(x,y,z) & \text{if } x^2 + y^2 > 1 \text{ and } 0 < z-5k < 1 \\ \beta(x,y,z) & \text{if otherwise.} \end{cases}$$

For example, consider a solution $\{x(t),y(t),z(t)\}$ such that $x(0) = y(0) = z(0) = 0$. Then, as long as $0 < z(t) < 2$, $z(t)$ increases and $\{x(t),y(t)\}$ spirals $(0,0)$ as t increases. If $2 < z(t) < 3$, we have $x'(t) = 16^2(3-z)(z-2)$, $y'(t) = 0$, $z'(t) = \beta(x,y,z)$, and

$$\frac{dx}{dz} = 16^2(3-z)(z-2)/\beta(x,y,z) \geq 16^2(3-z)(z-2).$$

Therefore, if t increases and $z(t_1) = 3$ at some t_1, then $x(t_1)>5$. Thus $\{x(t_1),y(t_1)\}$ cannot be in A_1. Since $z'(t) \leq 0$ for $x^2 + y^2 > 1$ and $5 < z < 6$ and since $x'(t) = y'(t) = 0$, the solution must approach some critical point. In other words, the solution is bounded. Generally, the same observation shows that any solution can enter A at most twice as t increases, and after that, the solution approaches some critical point since the system is recurrently defined. This shows the boundedness of solutions. Clearly, the points $(-4,0,0)$ and $(4,0,0)$ are attracting points.

Now we construct a periodic system of period ω. This idea was given by Yorke. Let $\lambda(t)$ be a continuous periodic function of period $\frac{\omega}{2}$ such that

$$\lambda(0) = \lambda(\tfrac{\omega}{2}) = 0, \ \lambda(t) > 0 \ \text{ for } \ t \ \varepsilon \ (0,\tfrac{\omega}{2}),$$

$$\int_0^{\omega/2} \lambda(t)dt = \pi,$$

and define a periodic system of period ω by

$$x' = f(t,x,y,z)$$
$$y' = g(t,x,y,z)$$
$$z' = h(t,x,y,z),$$

where

$$f(t,x,y,z) = \begin{cases} \lambda(t)f*(x,y,z) & (0 \le t \le \frac{\omega}{2}) \\[2mm] -\lambda(t)y & (\frac{\omega}{2} \le t \le \omega) \\[2mm] f(t-k\omega,x,y,z) & (k\omega \le t < (k+1)\omega), \end{cases}$$

$$g(t,x,y) = \begin{cases} \lambda(t)g*(x,y,z) & (0 \le t \le \frac{\omega}{2}) \\[2mm] \lambda(t)x & (\frac{\omega}{2} \le t \le \omega) \\[2mm] g(t-k\omega,x,y,z) & (k\omega \le t < (k+1)\omega) \end{cases}$$

and

$$h(t,x,y,z) = \begin{cases} \lambda(t)h*(x,y,z) & (0 \le t \le \frac{\omega}{2}) \\[2mm] 0 & (\frac{\omega}{2} \le t \le \omega) \\[2mm] h(t-k\omega,x,y,z) & (k\omega \le t < (k+1)\omega). \end{cases}$$

During the time change from $t = \frac{\omega}{2}$ to $t = \omega$, the (x,y)-plane is rotated about the origin, and hence, the point $(4,0,0)$ is transformed into $(-4,0,0)$. Therefore the solution through $(4,0,0)$ is a periodic solution of period 2ω, but not ω, and this solution is uniformly asymptotically stable. The only fixed point by this rotation is the origin, but the z-component of the solution through $(0,0,0)$ increases, and hence this solution cannot be a periodic solution.

16. Existence Theorems for Almost Periodic Solutions

Consider an almost periodic system

$$x' = f(t,x), \tag{16.1}$$

where $f(t,x) \in C(R \times S_{B*}, R^n)$, $S_{B*} = \{x;\ |x| < B*\}$, and $f(t,x)$ is
almost periodic in t uniformly for $x \in S_{B*}$. In the previous section, we have seen that the boundedness property of solutions of the
periodic system implies the existence of a periodic solution. However, for an almost periodic equation, the boundedness of solutions
does not necessarily imply the existence of an almost periodic solution even for scalar equations. Opial [57] has constructed an equation with all of its solutions bounded but none almost periodic. He considered a scalar equation $x' = f(t,x)$, where $f(t,x) \in C(R \times R, R)$ and
$f(t+1,x) = f(t,x+1) = f(t,x)$. Then f can be considered as a function on the surface of a torus. There exists a flow on the torus with
rotation number ρ which is irrational and hence the equation has no
periodic solutions (cf. [12], [28]). Then the equation
$y' = f(t,y+\rho t)-\rho = g(t,y)$ is an equation with no almost periodic
solution, while every solution is bounded. Fink and Frederickson [19],
by using Opial's equation, have constructed an almost periodic equation which has no almost periodic solutions, but the solutions are
uniformly ultimately bounded. Their equation is given by

$$y' = h(t,y) = \begin{cases} g(t,y) & \text{for } |y| \le 3 \\ g(t,y) + \dfrac{|y|}{y}C(y^2-9) & \text{for } |y| > 3, \end{cases}$$

where C is a constant such that $M+7C < 0$, $M = \sup|g(t,y)|$.

   Thus, in discussing the existence of an <u>almost periodic solu-
tion</u>, some kind of stability properties of a bounded solution has
been assumed. Miller [51] assumed that the bounded solution is
<u>totally stable</u> and Seifert [61] assumed the $\bar{\Sigma}$-stability of the
bounded solution, while Sell [64] assumed the <u>stability under disturbances from the hull</u>. All of them used the theory for dynamical
systems, and hence the uniqueness of solutions is assumed. These
results can be obtained by using the property of <u>asymptotically</u>

almost periodic functions without the uniqueness of solutions [14],
[83], [84], [85]. A basic theorem is the following due to Coppel [14].

Theorem 16.1. Suppose that system (16.1) has a bounded solu-
tion $\phi(t)$ defined on I such that $|\phi(t)| \leq B$, B < B*, for all
$t \geq 0$. If the solution $\phi(t)$ is asymptotically almost periodic, then
system (16.1) has an almost periodic solution.

Proof. Since $\phi(t)$ is asymptotically almost periodic, it has
the decomposition

$$\phi(t) = p(t) + q(t),$$

where $p(t)$ is almost periodic in t and $q(t)$ is continuous on
I and $q(t) \to 0$ as $t \to \infty$. Let $\{\tau_k\}$ be a sequence such that
$\tau_k \to \infty$ as $k \to \infty$ and $p(t+\tau_k) \to p(t)$ as $k \to \infty$. Then we have

$$\phi(t+\tau_k) = p(t+\tau_k)+q(t+\tau_k) \text{ for } t+\tau_k \geq 0.$$

Letting $k \to \infty$, we have $|p(t)| \leq B$ for all $t \in R$. By Theorem 2.7,
$f(t,p(t))$ is almost periodic in t. Since $\phi(t)$ is a solution of
(16.1),

$$\phi'(t) = f(t,p(t)) + f(t,\phi(t)) - f(t,p(t)).   \text{(16.2)}$$

It is clear that $f(t,\phi(t)) - f(t,p(t)) \to 0$ as $t \to \infty$. Thus (16.2)
shows that $\phi'(t)$ is also asymptotically almost periodic, and therefore, by Theorem 3.3,

$$p'(t) = f(t,p(t)) \text{ for } t \in R,   \text{(16.3)}$$

which shows that $p(t)$ is an almost periodic solution of (16.1).

Thus when an almost periodic system has an asymptotically almost periodic solution, we can always see the existence of an almost
periodic solution.

Theorem 16.2. If the bounded solution $\phi(t)$ of (16.1) is asymptotically almost periodic, then for any $g \in H(f)$ there exists a sequence $\{\tau_k\}$ such that $\phi(t+\tau_k)$ tends to an almost periodic solution of system

$$x' = g(t,x) \qquad\qquad (16.4)$$

uniformly on I as $k \to \infty$.

Proof. Since $\phi(t)$ is asymptotically almost periodic, we have $\phi(t) = p(t)+q(t)$, where $p(t)$ is almost periodic in t and $q(t) \to 0$ as $t \to \infty$. Since $g \in H(f)$, there exists a sequence $\{\tau_k\}$ such that $\tau_k \to \infty$ as $k \to \infty$, $f(t+\tau_k,x) \to g(t,x)$ uniformly on $R \times S_B$ and $p(t+\tau_k) \to p^*(t)$ uniformly on R as $k \to \infty$. Clearly $p^*(t)$ is almost periodic in t and $p^*(t)$ is a solution of (16.4).

As was seen in Section 12, the stability under disturbances from the hull is a sufficient condition for asymptotic almost periodicity. Therefore we have the following theorem.

Theorem 16.3. If system (16.1) has a solution $\phi(t)$ defined on I such that $|\phi(t)| \le B$, $B < B^*$, for all $t \ge 0$ and if $\phi(t)$ is stable under disturbances from $H(f)$ with respect to $K = \{x;\ |x| \le \frac{B^*+B}{2}\}$, then system (16.1) has an almost periodic solution $p(t)$ which is also stable under disturbances from $H(f)$ with respect to K.

Proof. The existence of an almost periodic solution $p(t)$ follows immediately from Theorems 12.4 and 16.1. As was seen in the proof of Theorem 16.1, we have $\phi(t) = p(t)+q(t)$, where $q(t) \to 0$ as $t \to \infty$. Let $\{\tau_k\}$ be a sequence, $\tau_k \to \infty$ as $k \to \infty$, such that $p(t+\tau_k) \to p(t)$ uniformly on R and $f(t+\tau_k,x) \to f(t,x)$ uniformly on $R \times K$ as $k \to \infty$. Since $\phi(t+\tau_k) = p(t+\tau_k)+q(t+\tau_k)$ and

$q(t) \rightarrow 0$ as $t \rightarrow \infty$, $\phi(t+\tau_k) \rightarrow p(t)$ uniformly on I. Therefore, by Theorem 13.1, the almost periodic solution $p(t)$ is stable under disturbances from $H(f)$ with respect to K.

Corollary 16.1. If the solution $\phi(t)$ in Theorem 16.3 is totally stable, then $\phi(t)$ is asymptotically almost periodic and system (16.1) has an almost periodic solution which is also totally stable.

This follows immediately from Theorems 12.3, 16.3, and 13.1.

Corollary 16.2. Assume that for every $g \varepsilon H(f)$, the solutions of

$$x' = g(t,x)$$

are unique for the initial condition. If the solution $\phi(t)$ in Theorem 16.3 is uniformly asymptotically stable, $\phi(t)$ is asymptotically almost periodic and consequently system (16.1) has an almost periodic solution $p(t)$ which is uniformly asymptotically stable.

This follows from Corollaries 14.2, 16.1, and Theorem 13.3.

Corollary 16.3. Assume that for every $(\psi,g) \varepsilon H(\phi,f)$, ψ is uniformly asymptotically stable with a common triple $(\delta(\cdot),\delta_0,T(\cdot))$. Then system (16.1) has an almost periodic solution which is uniformly asymptotically stable.

Proof. By Theorem 14.1 and Corollary 16.1, $\phi(t)$ is asymptotically almost periodic and system (16.1) has an almost periodic solution $p(t)$. Moreover, $\phi(t) = p(t) + q(t)$, where $q(t) \rightarrow 0$ as $t \rightarrow \infty$. For the sequence $\{\tau_k\}$, $\tau_k \rightarrow \infty$ as $k \rightarrow \infty$, such that $f(t+\tau_k,x) \rightarrow f(t,x)$ uniformly on $I \times S$ for any compact set S in S_{B*} and $p(t+\tau_k) \rightarrow p(t)$ uniformly on R as $k \rightarrow \infty$, we have

$$\phi(t+\tau_k) \rightarrow p(t) \quad \text{uniformly on I as } k \rightarrow \infty.$$

Since for every $(\psi, g) \in H(\phi, f)$, ψ is uniformly asymptotically stable, $p(t)$ must be uniformly asymptotically stable.

Corollary 16.4. Assume that for every $(\psi, g) \in H(\phi, f)$, ψ is unique for the initial condition. If the solution $\phi(t)$ is uniformly asymptotically stable, $\phi(t)$ is asymptotically almost periodic and consequently system (16.1) has an almost periodic solution.

The assumption implies the conclusion of Lemma 14.1, and hence we can prove this corollary by a method similar to the one in the proof of Theorem 14.1. For the details, see [83].

Now consider a periodic system

$$x' = f(t,x), \quad f(t+\omega, x) = f(t,x), \quad \omega > 0, \qquad (16.5)$$

where $f(t,x) \in C(R \times S_{B*}, R^n)$. As was seen in Section 15, the existence of a bounded solution of the periodic system does not necessarily imply the existence of a periodic solution of period ω. Deysach and Sell [16] have assumed that a bounded solution $\phi(t)$ is uniformly stable, and they have shown the existence of an almost periodic solution. In fact, as will be shown later, we cannot necessarily obtain a periodic solution. Sell [63] has assumed a bounded solution $\phi(t)$ to be weakly uniformly asymptotically stable, which is equivalent to uniformly asymptotic stability as was seen in Section 7, and has shown the existence of a periodic solution of period $m\omega$, $m \geq 1$. These results were obtained by using dynamical systems.

Let $\phi(t)$ be a solution of (16.5) defined on I such that $|\phi(t)| \leq B$, $B < B*$, for all $t \geq 0$. If $\phi(t)$ is uniformly stable, by Theorem 12.5, $\phi(t)$ is stable under disturbances from $H(f)$, and hence the following theorem follows from Theorem 16.3.

Theorem 16.4. If the solution $\phi(t)$ of the periodic system

(16.5) is uniformly stable, then $\phi(t)$ is asymptotically almost

periodic and system (16.5) has an almost periodic solution which is

also uniformly stable.

Proof. Since clearly $\phi(t)$ is asymptotically almost periodic,

$\phi(t)$ has the decomposition $\phi(t) = p(t) + q(t)$, where $p(t)$ is al-

most periodic and $q(t) \to 0$ as $t \to \infty$. Let $\tau_{k_j} = k_j\omega, k_j > 0$ in-

teger, such that $p(t+\tau_{k_j}) \to p^*(t)$ uniformly on R as $j \to \infty$. Then

$p^*(t)$ is almost periodic and $f(t+\tau_{k_j},x) = f(t,x)$. Therefore

$p(t+\tau_{k_j})$ is also a solution of (16.5), and thus $p^*(t)$ is an almost

periodic solution of (16.5). By Theorem 13.2, $p^*(t)$ is uniformly

stable.

Halanay [22 or 24] proved the first part of this theorem under

the assumption that solutions of (16.5) are unique for the initial

value problem.

Theorem 16.5. If the solution $\phi(t)$ of the periodic system

(16.5) is uniformly asymptotically stable, then system (16.5) has a

periodic solution of period $m\omega$ for some integer $m \geq 1$ which is

also uniformly asymptotically stable.

Proof. The existence of a periodic solution is a special case

of Theorem 15.10. However, we shall give a proof by using the

above result. Set $\phi_k(t) = \phi(t+k\omega)$, where $k > 0$ is an integer. By

Theorem 16.4, $\phi(t)$ is asymptotically almost periodic, and therefore a

subsequence $\{\phi_{k_j}(t)\}$ converges uniformly on I as $j \to \infty$. Since

$\phi_{k_j}(0)$ is convergent, there is an integer k_p such that

$$|\phi_{k_p}(0)-\phi_{k_{p+1}}(0)| < \delta_0,$$

where δ_0 is the number for uniformly asymptotic stability of $\phi(t)$.

Set $m = k_{p+1}-k_p$ and consider the solution $\phi(t+m\omega)$ of (16.5). Then we have

$$|\phi_m(k_p\omega)-\phi(k_p\omega)| = |\phi_{k_{p+1}}(0)-\phi_{k_p}(0)| < \delta_0,$$

and hence, we have

$$|\phi_m(t)-\phi(t)| \to 0 \quad \text{as} \quad t \to \infty. \tag{16.6}$$

On the other hand, we have $\phi(t) = p(t) + q(t)$, where $p(t)$ is almost periodic and $q(t) \to 0$ as $t \to \infty$. Therefore, by (16.6),

$$|p(t)-p(t+m\omega)| \to 0 \quad \text{as} \quad t \to \infty,$$

which implies that $p(t) = p(t+m\omega)$ for all $t \in R$, because $p(t)$ is almost periodic. This shows that system (16.5) has a periodic solution $p(t)$ of period $m\omega$.

Let τ_k be such that $\tau_k = km\omega$, where $k > 0$ is an integer. Then we have

$$\phi(t+km\omega) = p(t) + q(t+km\omega),$$

and hence $\phi(t+\tau_k) \to p(t)$ uniformly on I as $k \to \infty$. Thus, by Theorem 13.2, $p(t)$ is uniformly asymptotically stable.

The following example shows that for a periodic system, the existence of a bounded uniformly stable solution does not necessarily imply the existence of a subharmonic solution.

Example 16.1. Consider a periodic system of period 2π

$$
\begin{aligned}
x' &= (4-r^2)x-\pi y \\
y' &= (4-r^2)y+\pi x \\
z' &= \begin{cases} -(1+\sin t)z & (r > 1) \\ -r(1+\sin t)z + (1-r) & (r \leq 1) \end{cases} \\
w' &= \sin t
\end{aligned}
\tag{16.7}
$$

where $r = \sqrt{x^2+y^2}$.

Since $r' = r(4-r^2)$ and $\theta' = \pi$ in the polar coordinates on

(x,y)-plane, this system has a uniformly stable solution x =

2 cos πt, y = 2 sin πt, z = 0, w = -cos t and has no subharmonic

solution.

The last example in Section 15 shows that the existence of a

bounded uniformly asymptotically stable solution does not necessarily

imply the existence of a periodic solution of period ω.

17. <u>Separation Condition in Almost Periodic Systems</u>

Consider an almost periodic system

$$x' = f(t,x), \qquad\qquad (17.1)$$

where $f(t,x) \in C(R \times S_{B*}, R^n)$ and f(t,x) is almost periodic in t

uniformly for $x \in S_{B*}$. Let K be a compact set in S_{B*}. For sim-

plicity, if a solution φ(t) of (17.1) is in K for all $t \in (-\infty,\infty)$,

we say that φ is in K and denote this by $\phi \in K$. We shall discuss the

existence of an almost periodic solution under separation conditions.

One of these conditions was assumed by Amerio [1].

<u>Definition 17.1.</u> We say that the almost periodic system

(17.1) satisfies the <u>separation condition in K</u>, if for each $g \in H(f)$

there exists a λ(g) > 0 such that if x and y are distinct solu-

tions in K of

$$x' = g(t,x), \qquad\qquad (17.2)$$

then we have

$$|x(t)-y(t)| \geq \lambda(g) \quad \text{for all } t \in R. \qquad (17.3)$$

Remark. As will be seen, if system (17.1) satisfies the separation condition in K, we can choose a positive constant λ_0 inde-
pendent of g for which $|x(t)-y(t)| \geq \lambda_0$ for all t ε R. We shall
call λ_0 the separation constant in K.

In this section, we use Theorem 2.6 and we need many subsequences, and hence, to make the expressions simpler, we shall use the
following notations. For a sequence $\{\alpha_k\}$, we shall denote it by α
and $\beta \subset \alpha$ means that β is a subsequence of α. For $\alpha = \{\alpha_k\}$ and
$\beta = \{\beta_k\}$, $\alpha+\beta$ will denote the sequence $\{\alpha_k+\beta_k\}$. Moreover $T_\alpha x$ will
denote $\lim_{k\to\infty} x(t+\alpha_k)$, where $\alpha = \{\alpha_k\}$ and the limit exists for each t.
Now we discuss a result of Amerio for an almost periodic solu-
tion. For system (17.1), we assume that system (17.1) satisfies the
separation condition in K. Under this assumption, we have the following lemmas.

Lemma 17.1. For each g ε H(f), the number of solutions in K
is finite.

Proof. If there are an infinite number of solutions $x_k(t)$ in
K, there is a subsequence of $\{x_k(t)\}$ which tends to a solution x(t)
of (17.2) uniformly on any compact interval in R. Therefore we cannot have a constant $\lambda(g) > 0$.

Lemma 17.2. For all g ε H(f), we can choose a positive constant λ_0 independent of g for which

$$|x(t)-y(t)| \geq \lambda_0 \quad \text{for all} \quad t \ \varepsilon \ R. \qquad (17.4)$$

Proof. Let g_1 and g_2 be in H(f). Then there exists a
sequence $\{\gamma_k'\}$ such that

$$g_2(t,x) = \lim_{k\to\infty} g_1(t+\gamma_k',x)$$

uniformly on $R \times K$, that is, $T_\gamma g_1 = g_2$ uniformly on $R \times K$. Let $x_1(t)$ and $y_1(t)$ be solutions in K of the system

$$x' = g_1(t,x). \qquad (17.5)$$

Then there exists a subsequence $\gamma \subset \gamma'$ for which $T_\gamma x_1 = x_2$, $T_\gamma y_1 = y_2$ uniformly on any compact interval in R, and $x_2(t)$, $y_2(t)$ are solutions in K of

$$x' = g_2(t,x). \qquad (17.6)$$

If $x_1(t)$ and $y_1(t)$ are distinct solutions, we have

$$\inf_{t \in R} |x_1(t+\gamma_k) - y_1(t+\gamma_k)| = \inf_{t \in R} |x_1(t) - y_1(t)| = \alpha_{12} > 0,$$

and hence

$$\inf_{t \in R} |x_2(t) - y_2(t)| = \beta_{12} \geq \alpha_{12} > 0, \qquad (17.7)$$

which means that $x_2(t)$ and $y_2(t)$ are distinct solutions in K of system (17.6). Let $p_1 \geq 1$ and $p_2 \geq 1$ be the numbers of distinct solutions in K of (17.5) and (17.6), respectively. As was seen above, $x_2(t)$ and $y_2(t)$ were obtained from $x_1(t)$ and $y_1(t)$, and hence $p_1 \leq p_2$. In the same way, we have $p_2 \leq p_1$. Therefore we have $p_1 = p_2 = p$.

Now let

$$\alpha = \min\{\alpha_{ik};\ i,k = 1,\ldots,p,\ i \neq k\},$$
$$\beta = \min\{\beta_{jm};\ j,m = 1,\ldots,p,\ j \neq m\}.$$

Then by (17.7), we have $\alpha \leq \beta$. In the same way, we have $\alpha \geq \beta$. Therefore $\alpha = \beta = \lambda_0$.

Theorem 17.1. Suppose that system (17.1) satisfies the separation condition in K. If the system (17.1) has a solution $\phi(t)$ defined on I such that $\phi(t) \in K$ for $t \geq 0$, then $\phi(t)$ is asymptotically almost periodic and consequently system (17.1) has an almost periodic solution.

Proof. For any sequence $\{\tau_k'\}$ such that $\tau_k' \to \infty$ as $k \to \infty$, there is a subsequence $\{\tau_k\}$ such that $\phi(t+\tau_k)$ converges uniformly on any compact interval in I and $f(t+\tau_k, x)$ is uniformly convergent on $R \times K$. We shall show that $\phi(t+\tau_k)$ is convergent uniformly on I. Then $\phi(t)$ is asymptotically almost periodic and the existence of an almost periodic solution follows immediately from Theorem 16.1.

Suppose that $\phi(t+\tau_k)$ is not convergent uniformly on I. Then, for some $\varepsilon > 0$ such that $0 < \varepsilon < \frac{\lambda_0}{2}$, where λ_0 is the separation constant, there are sequences $\{t_j'\}$, $\{k_j\}$ and $\{m_j\}$ such that

$$k_j \to \infty, \ m_j \to \infty \quad \text{as} \quad j \to \infty,$$

$$|\phi(t_j'+\tau_{k_j}) - \phi(t_j'+\tau_{m_j})| \geq \varepsilon.$$

Since $\phi(\tau_k)$ is convergent, we have $|\phi(\tau_{k_j})-\phi(\tau_{m_j})| < \frac{\lambda_0}{2}$ if j is sufficiently large. Set $\psi_j(t) = \phi(t+\tau_{k_j})-\phi(t+\tau_{m_j})$. Then $|\psi_j(0)| < \frac{\lambda_0}{2}$ and $|\psi_j(t_j')| \geq \varepsilon$ for large j. Since $\varepsilon < \frac{\lambda_0}{2}$, there exists a $t_j \geq 0$ such that $c \leq |\psi_j(t_j)| < \frac{\lambda_0}{2}$. Thus we have sequences $\{t_j\}$, $\{\tau_{k_j}\}$ and $\{\tau_{m_j}\}$ for which

$$\varepsilon \leq |\phi(t_j+\tau_{k_j})-\phi(t_j+\tau_{m_j})| < \frac{\lambda_0}{2}. \tag{17.8}$$

Now we shall denote by γ the sequence $\{\tau_k\}$. Then $\gamma' = \{\tau_{k_j}\} \subset \gamma$ and $\gamma'' = \{\tau_{m_j}\} \subset \gamma$. Let $\alpha = \{t_j\}$. For the sequences α, γ' and γ'', by Theorem 2.6, there exist $\alpha' \subset \alpha$, $\beta \subset \gamma'$ and $\beta' \subset \gamma''$ such that

$$T_{\alpha'+\beta}f = T_{\alpha'}T_\beta f, \ T_{\alpha'+\beta'}f = T_{\alpha'}T_{\beta'}f \quad \text{exist uniformly on} \quad R$$

and

$T_{\alpha'+\beta}\phi = x$, $T_{\alpha'+\beta'}\phi = y$ exist uniformly on any compact interval in R.

Since $T_\beta f = T_{\beta'}f = g$, we have $T_{\alpha'+\beta}f = T_{\alpha'+\beta'}f = T_{\alpha'}g = h$. Thus x(t) and y(t) are solutions in K of the system

$$x' = h(t,x). \tag{17.9}$$

On the other hand, by (17.8), we have

$$\varepsilon \leq |x(0)-y(0)| \leq \frac{\lambda_0}{2},$$

which shows that x(t) and y(t) are distinct solutions in K of (17.9), and hence

$$|x(t)-y(t)| \geq \lambda_0 \quad \text{for}\quad t \in R.$$

Thus there arises a contradiction. Therefore $\phi(t+\tau_k)$ converges uniformly on I. This completes the proof.

Corollary 17.1 [1]. Suppose that system (17.1) satisfies the separation condition in K. If $\phi(t)$ is a solution of (17.1) such that $\phi(t) \in K$ for all $t \in R$, then $\phi(t)$ is almost periodic.

Proof. By Theorem 17.1, $\phi(t)$ is asymptotically almost periodic on I, and hence it has the decomposition $\phi(t) = p(t) + q(t)$, where p(t) is almost periodic and q(t) is continuous and $q(t) \to 0$ as $t \to \infty$. Since $\phi(t) \in K$, we have $p(t) \in K$ for all $t \in R$. If $\phi(t) \neq p(t)$ at some t, we have $|\phi(t)-p(t)| \geq \lambda_0 > 0$, where λ_0 is the separation constant. However $\phi(t)-p(t) \to 0$ as $t \to \infty$. This contradiction shows $\phi(t) \equiv p(t)$. This completes the proof.

Theorem 17.2. Suppose that $\xi(t)$ is the only almost periodic solution of (17.1) in K. Then, for any $g \in H(f)$ system (17.2) has only one almost periodic solution in K and its module is contained in the module of f.

Proof. Let $\{\tau_k\}$ be a sequence such that $f(t+\tau_k,x) \to g(t,x)$
uniformly on $R \times K$ and $\xi(t+\tau_k) \to \phi(t)$ uniformly on R as $k \to \infty$.
Clearly $\phi(t)$ is an almost periodic solution of (17.2) in K. Sup-
pose that $\psi(t)$ is also an almost periodic solution of (17.2) in K.
Let $\{t_k\}$ be a sequence such that $g(t+t_k,x) \to f(t,x)$ uniformly on
$R \times K$ as $k \to \infty$. Then there exists a subsequence $\{t_{k_j}\}$ of $\{t_k\}$
such that $\phi(t+t_{k_j})$ and $\psi(t+t_{k_j})$ tend to the almost periodic solution
$\xi(t)$ of (17.1) uniformly on R, because $\xi(t)$ is the only almost
periodic solution in K. If j is sufficiently large, we have
$|\phi(t+t_{k_j})-\psi(t+t_{k_j})| < \varepsilon$ for all $t \in R$. Letting $t = s-t_{k_j}$,
$|\phi(s)-\psi(s)| < \varepsilon$ for all $s \in R$. Since ε is arbitrary, we can see
that $\phi(t) \equiv \psi(t)$ for all $t \in R$. Let $\{\tau_k\}$ be any sequence such
that $f(t+\tau_k,x)$ is uniformly convergent on $R \times S$, S any compact
set in S_{B*}. If $\xi(t+\tau_k)$ is not uniformly convergent, we have two
almost periodic solutions in some equation in the hull by the same
argument in the proof of Theorem 17.1. This contradicts the uniqueness. Thus the module containment follows from Theorem 2.8.

Corollary 17.2. If for each $g \in H(f)$, system (17.2) has a
unique solution in K, then these solutions are almost periodic, the
modules of which are contained in the module of f.

In Theorem 17.1, we have seen a relationship between the separation condition and the asymptotically almost periodicity of a
bounded solution $\phi(t)$. Now we shall discuss some relationships between the separation condition and stability properties [54]. For a
system

$$x' = F(t,x), \qquad\qquad (17.10)$$

where $F \in C(R \times D, R^n)$, D an open set in R^n, and for a compact set

K in D, we shall denote by A(F,K) the family of solutions x(t)
of (17.10) such that x(t) ε K for all $t \geq t_0$ for some t_0 ε R.
For x ε A(F,K), let t_x be the infimun of t_0 such that x(t) ε K
for all $t \geq t_0$. t_x may be −∞. We denote by B(F,K) the set of
solutions x(t) of (17.10) such that x(t) ε K for all t ε R.

    Now we consider the almost periodic system (17.1). Let K
be a compact set in S_{B*}.

    Definition 17.2. x ε B(f,K) is said to be conditionally
totally stable in K, if for any ε > 0 there exists a δ(ε) > 0 such
that |x(t)−y(t,h)| < ε for all $t \geq t_0$ whenever y(t,h) ε A(f+h,K),
|x(t_0)−y(t_0,h)| < δ(ε) at some $t_0 > t_y$ and |h(t)| < δ(ε) on
[t_0, ∞) , where h(t) is continuous.

    Definition 17.3. The system (17.1) is said to be conditionally
totally stable in K, if every x ε B(f,K) is conditionally totally
stable in K.

    Remark. By the same argument as in the proof of Theorem 13.1,
we can see that conditionally total stability is inherited.

    Theorem 17.3. If the system (17.1) satisfies the separation
condition in K, then for each g ε H(f), system (17.2) is conditionally totally stable in K. Moreover, we can choose the number δ(·)
in Definition 17.2 so that δ(ε) depends only on ε and is independent of g and solutions.

    Proof. We shall prove that for any ε > 0 there exists a
δ(ε) > 0 such that for any g ε H(f) and any x ε B(g,K),
|x(t)−y(t)| < ε for all $t \geq t_0$ whenever y ε A(g+h,K),
|x(t_0)−y(t_0)| < δ(ε) for some $t_0 > t_y$ and |h(t)| < δ(ε) on
[t_0,∞). Suppose not. Then there exists an ε > 0 and sequences

$g_k \in H(f)$, $h_k(t)$, $x_k \in B(g_k, K)$, $y_k \in A(g_k + h_k, K)$, t_k and τ_k, $\tau_k > t_k$, such that

$$|h_k(t)| \leq \frac{1}{k} \quad \text{on} \quad t_k \leq t < \infty, \tag{17.11}$$

$$|x_k(t_k) - y_k(t_k)| < \frac{1}{k}, \quad t_k > t_{y_k}, \tag{17.12}$$

$$|x_k(\tau_k) - y_k(\tau_k)| = \varepsilon, \tag{17.13}$$

where we can assume that $\varepsilon \leq \dfrac{\lambda_0}{2}$ for the separation constant λ_0 in K.

Set $u_k(t) = x_k(t + \tau_k)$ and $v_k(t) = y_k(t + \tau_k)$. Then $u_k(t)$ and $v_k(t)$ are solutions of

$$x' = g_k(t + \tau_k, x)$$

and

$$x' = g_k(t + \tau_k, x) + h_k(t + \tau_k), \tag{17.14}$$

respectively, and $u_k(0) = x_k(\tau_k)$, $v_k(0) = y_k(\tau_k)$, and clearly

$$u_k(t) \in K \quad \text{for all} \quad t \in R$$

and

$$v_k(t) \in K \quad \text{for} \quad t \geq t_k - \tau_k \quad (t_k - \tau_k < 0).$$

Since $g_k(t + \tau_k, x) \in H(f)$ and $H(f)$ is compact by the uniform norm on $R \times K$, $\{g_k(t + \tau_k, x)\}$ has a subsequence, which we shall denote by $\{g_k(t + \tau_k, x)\}$ again, such that

$$g_k(t + \tau_k, x) \to p(t, x) \quad \text{uniformly on} \quad R \times K$$

as $k \to \infty$ and $p \in H(f)$. Also we can assume that $t_k - \tau_k$ tends to a τ as $k \to \infty$, where τ can be $-\infty$.

Since $\{v_k(t)\}$ is uniformly bounded and equicontinuous on any compact interval in (τ, ∞), there exists a subsequence of $\{v_k(t)\}$, which will be denoted by $\{v_k(t)\}$, and a function $\eta(t)$ defined on

(τ,∞) such that

$\qquad v_k(t) \to \eta(t)$ uniformly on any compact interval in (τ,∞)

as $k \to \infty$. Since $v_k(t)$ is a solution of (17.14) and $h_k(t+\tau_k) \to 0$

as $k \to \infty$, $\eta(t) \varepsilon A(p,K)$ and $\eta(t) \varepsilon K$ for $t > \tau$. By the same

argument, there exists a function $\xi(t)$ such that

$\qquad u_k(t) \to \xi(t)$ uniformly on any compact interval in R

and $\xi(t) \varepsilon B(p,K)$.

$\qquad$ If $\tau > -\infty$, $\eta(t) \varepsilon K$ for $t \geq \tau$ and $\lim\limits_{k\to\infty} v_k(t_k-\tau_k) = \eta(\tau)$.
Therefore,

$$|\xi(\tau)-\eta(\tau)| = \lim_{k\to\infty}|u_k(t_k-\tau_k)-v_k(t_k-\tau_k)|$$

$$= \lim_{k\to\infty}|x_k(t_k)-y_k(t_k)| = 0.$$

Thus we have a solution $\eta^* \varepsilon B(p,K)$, where

$$\eta^*(t) = \begin{cases} \eta(t) & \text{for } t \geq \tau \\ \xi(t) & \text{for } t < \tau. \end{cases}$$

If $\tau = -\infty$, we set $\eta^*(t) \equiv \eta(t) \varepsilon B(p,K)$. Thus we have two solutions
$\eta^*(t)$, $\xi(t)$ in $B(p,K)$. But

$$|\eta^*(0)-\xi(0)| = |\eta(0)-\xi(0)| = \lim_{k\to\infty}|v_k(0)-u_k(0)|$$

$$= \lim_{k\to\infty}|x_k(\tau_k)-y_k(\tau_k)| = \varepsilon > 0,$$

which shows that $\eta^*(t)$ and $\xi(t)$ are distinct solutions in $B(p,K)$.
Therefore $|\eta^*(t)-\xi(t)| \geq \lambda_0$ for all $t \varepsilon R$. However $|\eta^*(0)-\xi(0)| = \varepsilon < \dfrac{\lambda_0}{2}$. This contradiction proves the theorem.

$\qquad$ Theorem 17.4. Suppose that system (17.1) has a solution ϕ

in K which is conditionally totally stable. Then system (17.1) has an almost periodic solution.

    <u>Proof.</u> We shall show that $\phi(t)$ is asymptotically almost periodic. Then this will imply the existence of an almost periodic solution. Let $\{\tau_k\}$ be any sequence such that $\tau_k \to \infty$ as $k \to \infty$, and let $\phi_k(t) = \phi(t+\tau_k)$. Then $\phi_k(t)$ is a solution of

$$x' = f(t+\tau_k,x) \tag{17.15}$$

and clearly, $\phi_k(t)$ is conditionally totally stable in K with the same $\delta(\cdot)$ as for $\phi(t)$. We can assume that if $k_0(\epsilon)$ is sufficiently large and if $m \geq k \geq k_0(\epsilon)$,

$$|\phi_k(0)-\phi_m(0)| < \delta(\epsilon)$$

and

$$|f(t+\tau_k,x)-f(t+\tau_m,x)| < \delta(\epsilon) \quad \text{on} \quad R \times K.$$

Since $\phi_m(t) \in K$ for $t \geq 0$ and $\phi_m(t)$ is a solution of

$$x' = f(t+\tau_k,x) + f(t+\tau_m,\phi_m(t)) - f(t+\tau_k,\phi_m(t)),$$

conditionally total stability implies $|\phi_k(t)-\phi_m(t)| < \epsilon$ for all $t \geq 0$ if $m \geq k \geq k_0(\epsilon)$. This shows that $\phi(t)$ is asymptotically almost periodic. This completes the proof.

    <u>Remark.</u> Corollary 17.1 follows also from Theorem 17.3 and Theorem 17.4.

    <u>Definition 17.4.</u> $x \in B(f,K)$ is said to be <u>conditionally uniformly stable in</u> K, if for any $\epsilon > 0$ there exists a $\delta(\epsilon) > 0$ such that $|x(t)-y(t)| < \epsilon$ for all $t \geq t_0$ whenever $y \in A(f,K)$ and $|x(t_0)-y(t_0)| < \delta(\epsilon)$ at some $t_0 > t_y$. $x \in B(f,K)$ is said to be <u>conditionally uniformly asymptotically stable in</u> K, if x is conditionally uniformly stable in K and if there exists a $\delta_0 > 0$ and for any $\epsilon > 0$ there exists a $T(\epsilon) > 0$ such that

$|x(t)-y(t)| < \varepsilon$ for $t \geq t_0+T(\varepsilon)$ whenever $y \in A(f,K)$ and

$|x(t_0)-y(t_0)| < \delta_0$ at some $t_0 > t_y$.

Definition 17.5. The system (17.1) is said to be <u>conditionally</u> <u>uniformly asymptotically stable in K</u>, if every $x \in B(f,K)$ is conditionally uniformly asymptotically stable in K.

Theorem 17.5. The system (17.1) satisfies the separation condition in K if and only if for each $g \in H(f)$ system (17.2) is conditionally uniformly asymptotically stable in K with a common triple $(\delta_0, \delta(\cdot), T(\cdot))$.

Proof. Suppose that system (17.1) satisfies the separation condition in K. Then it follows from Theorem 17.3 that for any $\varepsilon > 0$ there exists a $\delta(\varepsilon) > 0$ such that for any $g \in H(f)$, every solution $x \in B(g,K)$ satisfies $|x(t)-y(t)| < \varepsilon$ for all $t \geq t_0$ whenever $y \in A(g,K)$ and $|x(t_0)-y(t_0)| < \delta(\varepsilon)$ at some $t_0 > t_y$. Now let δ_0 be a positive constant such that $\delta_0 < \delta(\frac{\lambda_0}{2})$, where λ_0 is the separation constant. We shall show that for any $\varepsilon > 0$ there is a $T(\varepsilon) > 0$ such that for each $g \in H(f)$, every solution $x \in B(g,K)$ satisfies

$$|x(t)-y(t)| < \varepsilon \quad \text{for all} \quad t \geq t_0+T(\varepsilon)$$

whenever $y \in A(g,K)$ and $|x(t_0)-y(t_0)| < \delta_0$ for some $t_0 > t_y$.

Suppose not. Then there exists an $\varepsilon > 0$ and sequences $g_k \in H(f)$, $x_k \in B(g_k,K)$, $y_k \in A(g_k,K)$, t_k, $t_k > t_{y_k}$, and τ_k, $\tau_k > t_k+k$, such that

$$|x_k(t_k)-y_k(t_k)| < \delta_0 (< \delta(\frac{\lambda_0}{2})) \qquad (17.16)$$

and

$$|x_k(\tau_k)-y_k(\tau_k)| \geq \varepsilon. \qquad (17.17)$$

Since (17.16) implies $|x_k(t)-y_k(t)| < \dfrac{\lambda_0}{2}$ for all $t \geq t_k$, by (17.17)

$$\varepsilon \leq |x_k(\tau_k)-y_k(\tau_k)| \leq \frac{\lambda_0}{2} . \qquad (17.18)$$

If we set $u_k(t) = x_k(t+\tau_k)$ and $v_k(t) = y_k(t+\tau_k)$, then $u_k(t)$ and $v_k(t)$ are solutions of

$$x' = g_k(t+\tau_k,x)$$

and $u_k(t) \in K$ for all $t \in R$, $v_k(t) \in K$ for $t \geq -k$. Thus we can assume that there exists an $h \in H(f)$ and $\xi \in B(h,K)$, $\eta \in B(h,K)$ such that

$$g_k(t+\tau_k,x) \to h(t,x) \quad \text{uniformly on} \quad R \times K$$

$$u_k(t) \to \xi(t), \ v_k(t) \to \eta(t) \quad \text{uniformly on any compact interval in } R$$

as $k \to \infty$. On the other hand, we have

$$|\xi(0)-\eta(0)| = \lim_{k\to\infty}|u_k(0)-v_k(0)| = \lim_{k\to\infty}|x_k(\tau_k)-y_k(\tau_k)| ,$$

which implies that, by (17.18),

$$\varepsilon \leq |\xi(0)-\eta(0)| \leq \frac{\lambda_0}{2} . \qquad (17.19)$$

Since $\xi \in B(h,K)$ and $\eta \in B(h,K)$, (17.19) contradicts the separation condition. This shows that for any $g \in H(f)$, system (17.2) is conditionally uniformly asymptotically stable in K.

Now assume that for each $g \in H(f)$ system (17.2) is conditionally uniformly asymptotically stable in K with a common triple $(\delta_0,\delta(\cdot),T(\cdot))$. First of all, we shall see that any two distinct solutions $x,y \in B(g,K)$, $g \in H(f)$, satisfy

$$\lim_{t\to-\infty} |x(t)-y(t)| \geq \delta_0. \qquad (17.20)$$

Suppose not. Then for some $g \in H(f)$, there exist two distinct solutions $x,y \in B(g,K)$ which satisfy

$$\lim_{t \to -\infty} |x(t)-y(t)| < \delta_0. \qquad (17.21)$$

Since $x(t) \not\equiv y(t)$, we have $|x(t_0)-y(t_0)| = \varepsilon$ at some t_0 and for
some ε . For this ε , let $T(\frac{\varepsilon}{2})$ be the number in Definition 17.4.
Then there is a t_1 such that $t_1 < t_0 - T(\frac{\varepsilon}{2})$ and

$$|x(t_1)-y(t_1)| < \delta_0$$

because of (17.21). The conditionally uniformly asymptotic stability
in K implies

$$|x(t_0)-y(t_0)| < \frac{\varepsilon}{2},$$

which contradicts $|x(t_0)-y(t_0)| = \varepsilon$. Thus we have (17.20).

Since K is a compact set, there are a finite number of cov-
erings which consist of m_0 balls with diameter $\frac{\delta_0}{4}$. We shall show
that the number of the solutions in $B(g,K)$ is at most m_0 . Suppose
not. Then there are m_0+1 solutions in K , $x_j(t)$, $j = 1,2,\ldots,m_0+1$,
and a t_2 such that

$$|x_i(t_2)-x_j(t_2)| \geq \frac{\delta_0}{2} \quad \text{for}\quad i \neq j, \qquad (17.22)$$

because we have (17.20). However, some two of these solutions, say
$x_i(t)$, $x_j(t)$ ($i \neq j$), are in one ball at time t_2 , and hence

$$|x_i(t_2)-x_j(t_2)| < \frac{\delta_0}{4} ,$$

which contradicts (17.22). Therefore the number of solutions in
$B(g,K)$ is $m \leq m_0$. Thus

$$B(g,K) = \{x_1(t),x_2(t),\ldots,x_m(t)\} \qquad (17.23)$$

and

$$\lim_{t \to -\infty} |x_i(t) - x_j(t)| \geq \delta_0, \quad i \neq j. \qquad (17.24)$$

Consider a sequence $\{\tau_k\}$ such that $\tau_k \to -\infty$ as $k \to \infty$ and

$$g(t+\tau_k, x) \to g(t,x) \quad \text{uniformly on} \quad R \times K$$

as $k \to \infty$. Since $x_j(t+\tau_k)$ is uniformly bounded and equicontinuous, $x_j(t+\tau_k)$ can be assumed to tend to a function $y_j(t)$ uniformly on any compact interval in R for j, $1 \leq j \leq m$. Clearly $y_j(t)$ is a solution in $B(g,K)$. Since we have

$$|y_i(t) - y_j(t)| = \lim_{k \to \infty} |x_i(t+\tau_k) - x_j(t+\tau_k)| \quad \text{for} \quad t \in R,$$

it follows from (17.24) that

$$|y_i(t) - y_j(t)| \geq \delta_0 \quad \text{for all} \quad t \in R \quad \text{and} \quad i \neq j. \qquad (17.25)$$

Since the number of solutions in $B(g,K)$ is m, $B(g,K)$ consists of $y_1(t), \ldots, y_m(t)$, and we have (17.25), which shows that system (17.1) satisfies the separation condition in K with the separation constant δ_0.

For an application of the separation condition, Seifert [62] assumed a bounded solution to be quasi uniformly asymptotically stable under an additional condition that for every $g \in H(f)$, solutions of (17.2) are unique, and he showed the existence of an almost periodic solution. Actually, these conditions imply the uniformly asymptotic stability of a bounded solution.

Lemma 17.3. Suppose that system (17.1) has a solution $\phi(t)$ defined on I such that $|\phi(t)| \leq B < B^*$ for all $t \geq 0$ and that for any $g \in H(f)$, solutions of (17.2) are unique for the initial value problem. If $\phi(t)$ is quasi uniformly asymptotically stable, that is, there exists a $\delta_0 > 0$ and for any $\varepsilon > 0$ and any $t_0 \in I$,

there exists a $T(\varepsilon) > 0$ such that $|\phi(t_0)-x_0| < \delta_0$ implies
$|\phi(t)-x(t,t_0,x_0)| < \varepsilon$ for all $t \geq t_0+T(\varepsilon)$. Then $\phi(t)$ is uniformly
asymptotically stable.

 Proof. Since for any $g \in H(f)$, solutions of (17.2) are unique,
by Lemma 14.1, for any ε such that $0 < \varepsilon < B^*-B$ there is a
$\delta(\varepsilon) > 0$, $\delta(\varepsilon) < \delta_0$, such that for any $t_0 \in I$, $|\phi(t_0)-x_0| < \delta(\varepsilon)$ implies

$$|\phi(t)-x(t,t_0,x_0)| < \varepsilon \text{ on } [t_0,t_0+T(\varepsilon)].$$

Therefore, if $|\phi(t_0)-x_0| < \delta(\varepsilon)$, then $|\phi(t)-x(t,t_0,x_0)| < \varepsilon$ for all
$t \geq t_0$. This proves the uniform stability of $\phi(t)$, and hence $\phi(t)$
is uniformly asymptotically stable.

18. Uniform Stability and Existence of Almost Periodic Solutions

As was seen in Theorem 16.4, if a periodic system has a bounded
uniformly stable solution, then there exists an almost periodic solution, and Example 16.1 shows that there does not necessarily exist a
subharmonic solution. The system (16.7) has a quasi-periodic solution,
and thus the module of the almost periodic solution is not contained
in the module of the system. This shows that for an almost periodic
system, uniform stability and also stability under disturbances from
the hull do not give the module containment.

In this section, we shall discuss the case where an almost
periodic system has a bounded uniformly stable solution. Consider
an almost periodic system

$$x' = f(t,x), \tag{18.1}$$

where $f(t,x) \in C(R \times S_{B^*}, R^n)$ and $f(t,x)$ is almost periodic in t
uniformly for $x \in S_{B^*}$. The results in this section are due to Fink
[18] and Nakajima.

Let K be a compact set in S_{B^*}. First of all, we shall prove the following lemma which gives a condition for asymptotic almost periodicity. The condition in the lemma is also a necessary condition. See [18].

Lemma 18.1. Let $f(t)$ be a continuous function on $[0,\infty)$. If given a sequence $\gamma' = \{\gamma_k'\}$ such that $\gamma_k' \to \infty$ as $k \to \infty$, there exists a subsequence $\gamma \subset \gamma'$ and a number $\lambda(\gamma) > 0$ such that $T_\gamma f$ exists pointwise and if α is a sequence with $\alpha_k > 0$ and $\beta' \subset \gamma$, $\beta'' \subset \gamma$ are such that $T_{\alpha+\beta'}f = g$ and $T_{\alpha+\beta''}f = h$ exist pointwise, either $g \equiv h$ or $|g(t)-h(t)| \geq 2\lambda(\gamma)$ for all $t \in [0,\infty)$, then $f(t)$ is asymptotically almost periodic.

Proof. Let γ' be a sequence such that $\gamma_k' \to \infty$ as $k \to \infty$. We shall show that there exists a subsequence $\gamma \subset \gamma'$ such that $T_\gamma f$ exists uniformly on I. Then $f(t)$ is asymptotically almost periodic. By the hypothesis, there exists a subsequence $\gamma \subset \gamma'$ such that $T_\gamma f$ exists pointwise. Suppose that $f(t+\gamma_k)$ is not uniformly convergent on I. Then there exist sequences τ' with $\tau_k' > 0$, $\alpha' \subset \gamma$, $\beta' \subset \gamma$ and $\varepsilon > 0$ such that

$$|f(\tau_k'+\alpha_k') - f(\tau_k'+\beta_k')| \geq \varepsilon,$$

where we can assume that $\varepsilon < \lambda(\gamma)$. Since $T_\gamma f(0)$ exists, we have

$$|f(\alpha_k')-f(\beta_k')| < \lambda(\gamma)$$

for large k. Therefore, letting $m(t) = f(t+\alpha_k')-f(t+\beta_k')$, we have $|m(0)| < \lambda(\gamma)$ and $|m(\tau_k')| \geq \varepsilon$ for large k. Thus there exists τ_k'' such that $\varepsilon \leq |m(\tau_k'')| < \lambda(\gamma)$. Consider the sequences $\alpha'+\tau''$ and $\beta'+\tau''$. By the hypothesis, there exist subsequences $\alpha+\tau \subset \alpha'+\tau''$ and $\beta+\tau \subset \beta'+\tau''$ such that $T_{\alpha+\tau}f = g$ and $T_{\beta+\tau}f = h$ exist pointwise and $g \equiv h$ or $|g(t)-h(t)| \geq 2\lambda(\gamma)$ on I. However,

$$|g(0)-h(0)| = \lim_{k\to\infty}|f(\alpha_k+\tau_k)-f(\beta_k+\tau_k)| \geq \varepsilon.$$

Therefore $0 < \varepsilon \leq |g(0)-h(0)| \leq \lambda(\gamma)$, which contradicts

$|g(t)-h(t)| \geq 2\lambda(\gamma)$. This shows that $f(t+\gamma_k)$ converges uniformly

on I.

   Remark. Clearly, if f(t) is continuous on $(-\infty,0]$ and if

given a sequence $\{\gamma_k'\}$ such that $\gamma_k' \to -\infty$ as $k \to \infty$, the condition

in the lemma is satisfied, f(t) is <u>asymptotically almost periodic on</u>

$(-\infty,0]$, that is,

$$f(t) = p(t)+q(t),$$

where p(t) is almost periodic and q(t) is continuous on $(-\infty,0]$

and $q(t) \to 0$ as $t \to -\infty$.

   <u>Definition 18.1.</u> A property P of a solution ϕ in K of

(18.1) is said to be a <u>semiseparating property</u>, if for any other solu-

tion ψ in K of (18.1) which has property P, there exists a

$\lambda(\phi,\psi) > 0$ such that $|\phi(t)-\psi(t)| \geq \lambda(\phi,\psi)$ for $t \varepsilon (-\infty,0]$.

   <u>Lemma 18.2.</u> Uniform stability is a semiseparating property

if the solution is unique for the initial value problem.

   Proof. Let ϕ be a uniformly stable solution in K and let

ψ be any other solution in K. Then $|\phi(0)-\psi(0)| = \varepsilon > 0$ for some

ε, because of uniqueness. If there is a $t_0 \varepsilon (-\infty,0)$ such that

$|\phi(t_0)-\psi(t_0)| < \delta(\frac{\varepsilon}{2})$, where $\delta(\cdot)$ is the number for the uniform sta-

bility of $\phi(t)$, then $|\phi(0)-\psi(0)| < \frac{\varepsilon}{2}$, which is a contradiction.

Thus $|\phi(t)-\psi(t)| \geq \delta(\frac{\varepsilon}{2})$ for all $t \varepsilon (-\infty,0]$. This proves the lemma.

   <u>Lemma 18.3.</u> Suppose that the property P is inherited and is

semiseparating. If system (18.1) has only a finite number of solu-

tions in K with property P, then every equation in H(f)

$$x' = g(t,x) \qquad (18.2)$$

has the same number of solutions in K with property P and the separation constant $\lambda(\phi,\psi)$ can be chosen independently of solutions and equations.

     <u>Proof</u>. Since system (18.1) has only a finite number of solutions in K with property P, the separation constant can be assumed to depend only on f. We denote it by $\lambda(f)$. Let ϕ and ψ be solutions of (18.1) in K with property P and $g \in H(f)$. Then there exists a sequence $\alpha = \{\alpha_k\}$ such that $\alpha_k \to -\infty$ as $k \to \infty$, $f(t+\alpha_k,x) \to g(t,x)$ uniformly on $R \times K$, $\phi(t+\alpha_k) \to \xi(t)$ and $\psi(t+\alpha_k) \to \eta(t)$ uniformly on any compact interval in R. Clearly $\xi(t)$ and $\eta(t)$ are solutions of (18.2) in K with property P, and also $|\xi(t)-\eta(t)| \geq \lambda(f)$. Thus, if $\phi_1,\ldots,\phi_k$ are the solutions of (18.1) in K with property P, then $T_\alpha\phi_1,\ldots,T_\alpha\phi_k$ are distinct solutions of (18.2) in K with property P. Therefore system (18.2) has at least k such solutions. On the other hand, if $\psi_1,\ldots,\psi_m$ are solutions of (18.2) in K with property P, by a similar construction $T_\beta\psi_1,\ldots,T_\beta\psi_m$, where $T_\beta g = f$, are solutions of (18.1) in K with property P. Therefore $m \leq k$, but $m \geq k$. Thus $m = k$ and we can choose a separation constant λ_0 independent of solutions and equations.

     Now we prove the following theorem for the existence of an almost periodic solution.

     <u>Theorem 18.1</u>. Let P be an inherited property which is semi-separating. If system (18.1) has only a finite number of solutions in K with property P, then every such solution is asymptotically almost periodic on $(-\infty,0]$ and there exists an almost periodic solution in K.

Proof. Let ϕ be a solution in K with property P and let λ_0 be the separation constant. Note that we can assume λ_0 is independent of solutions and equations by Lemma 18.3. Now we shall show that ϕ satisfies the condition in Lemma 18.1 with I replaced by $(-\infty,0]$ and $\lambda = \dfrac{\lambda_0}{2}$. Let γ_k' be a sequence such that $\gamma_k' \to -\infty$ as $k \to \infty$. Then there exists a subsequence $\gamma \subset \gamma'$ such that

$$T_\gamma f = g \quad \text{uniformly on} \quad R \times K$$

and

$$T_\gamma \phi \quad \text{exists uniformly on any compact set on} \quad R.$$

Let $\alpha = \{\alpha_k\}$, $\alpha_k < 0$, $\beta' \subset \gamma$ and $\beta'' \subset \gamma$ such that $T_{\alpha+\beta'}\phi = \xi$ and $T_{\alpha+\beta''}\phi = \eta$ exist. We can assume that ξ and η are solutions in K with property P of $x' = h(t,x)$, where

$$h = T_\alpha g = T_\alpha T_\gamma f.$$

Therefore $\xi \equiv \eta$ or $|\xi(t)-\eta(t)| \geq \lambda_0 = 2\lambda$ on $(-\infty,0]$. Therefore ϕ is asymptotically almost periodic on $(-\infty,0]$. This completes the proof.

The following corollary follows immediately from Lemma 18.2 and Theorems 13.3 and 18.1.

Corollary 18.1. Suppose that for each $g \; \epsilon \; H(f)$, solutions of (18.2) are unique for the initial value problem. If system (18.1) has a finite number of solutions in K which are uniformly stable, then each such solution is asymptotically almost periodic on $(-\infty,0]$ and there is an almost periodic solution in K.

The special case when there is only one uniformly stable solution is a stronger version of a result in [20] where it is required that this uniformly stable solution is the only solution in K.

We now consider a linear system

$$x' = A(t)x + f(t), \qquad (18.3)$$

where A(t) is an almost periodic n × n matrix function defined on R and f(t) is an almost periodic function defined on R. Corresponding to system (18.3), consider the homogeneous linear system

$$x' = A(t)x \qquad (18.4)$$

and the equation in the hull

$$x' = B(t)x, \quad B \in H(A). \qquad (18.5)$$

If system (18.3) has a bounded solution defined on I which is uniformly stable, the zero solution of (18.4) and also the zero solution of (18.5) are uniformly stable. In this case, system (18.3) has an almost periodic solution. This follows from a result of Favard. Favard proved the following theorem [17].

Theorem 18.2. Suppose that for every B ∈ H(A), every non-trivial bounded solution x(t) of (18.5) on R satisfies $\inf_{t \in R} |x(t)| > 0$. If system (18.3) has a solution bounded on I, then there exists an almost periodic solution p(t) of (18.3) such that m(p) ⊂ m(A,f).

Proof. Since system (18.3) has a solution bounded on I, there exists a solution ψ(t) of (18.3) which is defined on R and is bounded by some constant B, that is, $|\psi(t)| \leq B$ for all t ∈ R. In this proof, we use the Euclidean norm and for a bounded solution x = x(t) on R, let $|x| = \sup_{t \in R} |x(t)|$. First of all, we shall see that there exists a solution φ(t) on R such that $|\phi|$ is minimal. Such a solution is called a minimal solution of (18.3). Also we see that the minimal solution is unique.

Let $K = \{x; |x| \leq B+1\}$, and let $\lambda = \inf\{|x|; x(t) \in K$ for all $t \in R$ and $x(t)$ is a solution of (18.3)$\}$. Since $A(t)x+f(t)$ is bounded on $R \times K$ and $\psi(t) \in K$ for all $t \in R$, we can easily see that there exists a solution ϕ such that $|\phi| = \lambda$, that is, ϕ is a minimal solution. Now suppose that x and y are distinct solutions of (18.3) such that $|x| = |y| = \lambda$. Then $\frac{1}{2}(x(t)+y(t))$ is a solution of (18.3) and $\frac{1}{2}(x(t)-y(t))$ is a nontrivial bounded solution of (18.4). Therefore $|\frac{1}{2}(x(t)-y(t))| \geq \delta > 0$ for all $t \in R$. Since we have

$$\left|\frac{x(t)+y(t)}{2}\right|^2, \quad \left|\frac{x(t)-y(t)}{2}\right|^2 = \frac{|x(t)|^2+|y(t)|^2}{2} \leq \lambda^2,$$

$|\frac{x(t)+y(t)}{2}|^2 \leq \lambda^2-\delta^2$ and hence $|\frac{x+y}{2}| < \lambda$ This contradicts that $|\phi|$ is minimal.

Thus we see that for each $(B,g) \in H(A,f)$, system

$$x' = B(t)x + g(t) \tag{18.6}$$

has a unique minimal solution. Moreover, it is easily seen that the minimal solution is inherited and λ is the common number for any equation (18.6).

Now let $p(t)$ be the minimal solution of (18.3) and let $\{\tau_k\}$ be a sequence such that $A(t+\tau_k)x+f(t+\tau_k) \to B(t)x+g(t)$ uniformly on $R \times K$ as $k \to \infty$. Suppose $p(t+\tau_k)$ is not uniformly convergent on R. Then, by the same argument in the proof of Theorem 17.1, we can find two minimal solutions of some system in the hull. This contradicts the uniqueness of the minimal solution. Thus we see that $p(t)$ is an almost periodic solution of (18.3) and $m(p) \subset m(A,f)$.

Theorem 18.3.   If the almost periodic linear system (18.3) has a bounded solution on I which is uniformly stable, then the system (18.3) has an almost periodic solution $p(t)$ and $m(p) \subset m(A,f)$.

Proof.   We shall show that if system (18.3) has a bounded solution on I which is uniformly stable, the assumption in Theorem

18.2 is satisfied. Then the conclusion follows from Theorem 18.2.

For $B \in H(A)$, let $\{\phi_j(t)\}$, $j = 1,2,\ldots,m$, $m \leq n$, be a basis of the

space of solutions of (18.5) which are bounded on R. Let $\{\tau_k\}$ be a

sequence such that $\tau_k \to -\infty$ as $k \to \infty$ and $B(t+\tau_k) \to B(t)$ uniformly

on R as $k \to \infty$. Moreover, we can assume that $\phi_j(t+\tau_k) \to \psi_j(t)$ uni-

formly on any compact interval in R for all $j = 1,\ldots,m$ as $k \to \infty$.

Then $\psi_j(t)$ is a solution of (18.5) which is bounded on R.

   Now we show that $\{\psi_j(t)\}$ are linearly independent. Suppose

$\sum_{j=1}^{m} c_j \psi_j(0) = 0$ for some constants c_j. Then $x(t) = \sum_{j=1}^{m} c_j \phi_j(t)$ is

a solution of (18.5) and $x(\tau_k) \to 0$ as $k \to \infty$. Since the zero solu-

tion of (18.5) is also uniformly stable, $x(t) \equiv 0$. Therefore $c_j = 0$,

$j = 1,2,\ldots,m$, because $\{\phi_j(t)\}$ are linearly independent. Thus

$\{\psi_j(t)\}$ are also linearly independent bounded solutions of (18.5) and

hence it forms a basis.

   Let $x(t)$ be a nontrivial bounded solution on R of (18.5).

Then we have

$$x(t) = \sum_{j=1}^{m} \lambda_j \psi_j(t)$$

for some constants λ_j. Since $x(0) \neq 0$, there exists a k_0 such

that $\sum_{j=1}^{m} \lambda_j \phi_j(\tau_k) \neq 0$ for $k \geq k_0$. Since the zero solution of (18.5)

is uniformly stable, we can find a constant $\delta > 0$ such that

$$\left| \sum_{j=1}^{m} \lambda_j \phi_j(t) \right| \geq \delta \quad \text{for} \quad t \leq \tau_{k_0}.$$

Therefore, for any $t \in R$

$$\left| \sum_{j=1}^{m} \lambda_j \psi_j(t) \right| = \lim_{k \to \infty} \left| \sum_{j=1}^{m} \lambda_j \phi_j(t+\tau_k) \right| \geq \delta$$

or

$$|x(t)| \geq \delta \quad \text{for all} \quad t \in R.$$

This shows that the assumption in Theorem 18.2 is satisfied.

Remark. Since the zero solution of (18.4) is uniformly stable, the almost periodic solution p(t) is uniformly stable. The case where the zero solution of (18.4) is uniformly asymptotically stable will be discussed in Section 19.

19. Existence of Almost Periodic Solutions By Liapunov Functions

Consider an almost periodic system

$$x' = f(t,x),\qquad\qquad (19.1)$$

where $f(t,x) \in C(R \times S_{B*}, R^n)$, $S_{B*} = \{x;\ |x| < B*\}$, and $f(t,x)$ is almost periodic in t uniformly for $x \in S_{B*}$. Now, by using Liapunov functions, we discuss the existence of an almost periodic solution which is uniformly asymptotically stable in the whole, that is, every solution which remains in S_{B*} in the future approaches the almost periodic solution as $t \to \infty$. To discuss this, corresponding to system (19.1), we consider the system

$$x' = f(t,x),\quad y' = f(t,y).\qquad\qquad (19.2)$$

Theorem 19.1. Suppose that there exists a Liapunov function $V(t,x,y)$ defined on $0 \le t < \infty$, $|x| < B*$, $|y| < B*$ which satisfies the following conditions;

(i) $a(|x-y|) \le V(t,x,y) \le b(|x-y|)$, where $a(r)$ and $b(r)$ are continuous, increasing and positive definite,

(ii) $|V(t,x_1,y_1)-V(t,x_2,y_2)| \le K\{|x_1-x_2 + y_1-y_2|\}$, where $K > 0$ is a constant,

(iii) $\dot{V}_{(19.2)}(t,x,y) \le -\alpha V(t,x,y)$, where $\alpha > 0$ is a constant.

Moreover, suppose that there exists a solution $\phi(t)$ of (19.1) such that $|\phi(t)| \leq B < B^*$ for $t \geq 0$. Then, in the region $R \times S_{B^*}$, there exists a unique uniformly asymptotically stable almost periodic solution p of (19.1) which is bounded by B, and $m(p) \subset m(f)$. In particular, if $f(t,x)$ is periodic in t of period ω, then there exists a unique uniformly asymptotically stable periodic solution of (19.1) of period ω.

Proof. Let $\{\tau_k\}$ be a sequence such that $\tau_k \to \infty$ as $k \to \infty$. Set $\phi_k(t) = \phi(t+\tau_k)$. Then $\phi_k(t)$ is a solution of $x' = f(t+\tau_k,x)$ through $(0,\phi(\tau_k))$. Since $f(t,x)$ is almost periodic, there exists a subsequence of $\{\tau_k\}$, which we shall denote by $\{\tau_k\}$ again, such that $f(t+\tau_k,x)$ converges uniformly on $R \times \bar{S}_B$ as $k \to \infty$. For a given $\varepsilon > 0$, choose an integer $k_0(\varepsilon)$ so large that if $m \geq k \geq k_0(\varepsilon)$,

$$b(2B)e^{-\alpha\tau_k} < \frac{a(\varepsilon)}{2} \tag{19.3}$$

and

$$|f(t+\tau_k,x)-f(t+\tau_m,x)| < \frac{a(\varepsilon)\alpha}{2K} \quad \text{on} \quad R \times \bar{S}_B. \tag{19.4}$$

From conditions (ii) and (iii), it follows that

$$V'(t,\phi(t),\phi(t+\tau_m-\tau_k)) \leq -\alpha V(t,\phi(t),\phi(t+\tau_m-\tau_k))$$

$$+ K|f(t+\tau_m-\tau_k,\phi(t+\tau_m-\tau_k))-f(t,\phi(t+\tau_m-\tau_k))|,$$

because $\phi(t+\tau_m-\tau_k)$ is a solution of

$$x' = f(t+\tau_m-\tau_k,x).$$

By (19.4), we have

$$V'(t,\phi(t),\phi(t+\tau_m-\tau_k)) \leq -\alpha V(t,\phi(t),\phi(t+\tau_m-\tau_k)) + \frac{a(\varepsilon)\alpha}{2},$$

which implies that

$$V(t+\tau_k,\phi(t+\tau_k),\phi(t+\tau_m)) \leq e^{-\alpha(t+\tau_k)} V(0,\phi(0),\phi(\tau_m-\tau_k)) + \frac{a(\varepsilon)}{2}.$$

Thus, if $m \geq k \geq k_0(\varepsilon)$, by (19.3),

$$v(t+\tau_k,\phi(t+\tau_k),\phi(t+\tau_m)) < a(\varepsilon).$$

Therefore, by (i), we have

$$|\phi(t+\tau_k)-\phi(t+\tau_m)| < \varepsilon \quad \text{for all} \quad t \geq 0 \quad \text{if} \quad m \geq k \geq k_0,$$

which shows that $\phi(t)$ is asymptotically almost periodic, and hence
system (19.1) has an almost periodic solution $p(t)$ which is bounded
by B. By using the Liapunov function $V(t,x,y)$, we can easily see
that $p(t)$ is uniformly asymptotically stable and every solution re-
maining in S_{B*} approaches $p(t)$ as $t \to \infty$, which implies the uniqueness of $p(t)$. This also implies that $m(p) \subset m(f)$ by Theorem
17.2.

In the case where $f(t,x)$ is periodic in t of period ω ,
$p(t+\omega)$ is also a solution of (19.1) which remains in S_{B*} , and hence
$p(t+\omega) \to p(t)$ as $t \to \infty$. Thus we have $p(t+\omega) = p(t)$. This completes
the proof.

This theorem can be proved also in the following way. If we
set

$$W(t,x) = V(t,x,\phi(t)),$$

then $W(t,x)$ is defined for $t \geq 0$, $|x-\phi(t)| < \frac{B*-B}{2}$ and satisfies

(iv) $a(|x-\phi(t)|) \leq W(t,x) \leq b(|x-\phi(t)|)$

(v) $|W(t,x)-W(t,y)| \leq K|x-y|$

(vi) $\dot{W}_{(19.1)}(t,x) \leq -\alpha W(t,x).$

By using $W(t,x)$, we can show that $\phi(t)$ is integrally asymptotically
stable, and hence, by Theorem 12.1, $\phi(t)$ is totally stable. There-

fore, by Corollary 16.1, there exists an almost periodic solution of (19.1).

In Theorem 19.1, condition (iii) can be replaced by

(iii)' $\dot{V}_{(19.2)}(t,x,y) \leq -c(|x-y|)$, where $c(r)$ is continuous and positive definite.

In the case, condition (vi) becomes

(vi)' $\dot{W}_{(19.1)}(t,x) \leq -c(|x-\phi(t)|)$,

and hence, we can see that $\phi(t)$ is totally stable.

The following theorem follows immediately from the above result.

Theorem 19.2. Suppose that $f(t,x)$ is defined on $R \times R^n$ and is almost periodic in t uniformly for $x \in R^n$ and that the solutions of (19.1) are ultimately bounded for bound $B > 0$. Moreover, assume that there exists a Liapunov function $V(t,x,y)$ defined on $I \times S_{B*} \times S_{B*}$, $B < B*$, which satisfies the following conditions;

(i) $a(|x-y|) \leq V(t,x,y) \leq v(|x-y|)$, where $a(r)$ and $b(r)$ are continuous, increasing and positive definite.

(ii) $|V(t,x_1,y_1)-V(t,x_2,y_2)| \leq K\{|x_1-x_2|+|y_1-y_2|\}$,

(iii) $\dot{V}_{(19.2)}(t,x,y) \leq -c(|x-y|)$, where $c(r)$ is continuous and positive definite.

Then system (19.1) has a unique almost periodic solution which is uniformly asymptotically stable in the large. Furthermore, the module of this solution is contained in the module of $f(t,x)$.

For related results, see [37], [59], [81].

Example 19.1. Consider a second order differential equation

$$x'' + kf(x)x'+x = kp(t), \quad k > 0, \qquad (19.5)$$

where we assume that $f(x) > 0$ is continuous and $F(x) = \int_0^x f(u)du \to \pm\infty$ as $x \to \pm\infty$, respectively, and $p(t)$ is almost periodic and $P(t) = \int_0^t p(s)ds$ is bounded. Note that $P(t)$ is also almost periodic. Then equation (19.5) has an almost periodic solution which is uniformly asymptotically stable in the large.

To see this, consider an equivalent system and its associated system

$$x' = y-kF(x) + kP(t), \quad y' = -x \tag{19.6}$$

and

$$\begin{cases} x' = y-kF(x) + kP(t), \quad y' = -x \\ u' = v-kF(u) + kP(t), \quad v' = -u. \end{cases} \tag{19.7}$$

Under the assumptions, we can see that the solutions of (19.6) are uniformly ultimately bounded for some constant B (cf. [80]). For B^*, $B^* > B$, consider a Liapunov function

$$V(t,x,y,u,v) = (x-u)^2-2c(x-u)(y-v)+(y-v)^2,$$

where $c > 0$ is a small constant. Then this Liapunov function satisfies conditions (i) and (ii) in Theorem 19.2. $\dot{V}_{(19.7)}(t,x,y,u,v)$ will satisfy also condition (iii). In fact,

$$\dot{V}_{(19.7)}(t,x,y,u,v)$$
$$= 2(x-u)\{y-kF(x)-v+kF(u)\}-2c\{y-kF(x)-v+kF(u)\}(y-v)$$
$$\quad + 2c(x-u)^2-2(y-v)(x-u)$$

$$= -\{2k\frac{F(x)-F(u)}{x-u}-2c\}(x-u)^2+2ck\frac{F(x)-F(u)}{x-u}(x-u)(y-v)-2c(y-v)^2.$$

Since $f(x) > 0$, there are $M > 0$ and $N > 0$ such that $N \geq 2k\dfrac{F(x)-F(u)}{x-u} \geq M$ for $|x| < B^*$ and $|u| < B^*$. Therefore we have

$$\dot{V}_{(19.7)}(t,x,y,u,v) \leq -(M-2c)(x-u)^2+cN|x-u||y-v|-2c(y-v)^2.$$

Thus, if $c^2N^2 < 8c(M-2c)$ or $c < \dfrac{2M}{N^2+16}$, $\dot{V}_{(19.7)}(t,x,y,u,v)$ satis-

fies condition (iii) in Theorem 19.2. Therefore an almost periodic

solution $x = p(t)$, $y = q(t)$ of (19.6) is uniformly asymptotically

stable in the large. Moreover, since we have

$$p'(t) = q(t)-kF(p(t))+kP(t),$$

the derivative of the almost periodic solution is also almost periodic.

   Theorem 19.3. Suppose that there exists a Liapunov function

$V(t,x,y)$ defined on $I \times S_{B*} \times S_{B*}$ which satisfies the conditions

(i), (iii) in Theorem 19.1 and

   (ii)' $|V(t,x_1,y_1)-V(t,x_2,y_2)| \leq K|(x_1-x_2)-(y_1-y_2)|$, where
          $K > 0$ is a constant.

Moreover, suppose that system (19.1) has a solution $\phi(t)$ such that

$|\phi(t)| \leq c$ for $t \geq 0$ and some constant $c > 0$, $c < B*$. Consider a

system

$$x' = f(t,x) + g(t), \tag{19.8}$$

where $g(t)$ is almost periodic in t and $|g(t)| \leq R$ for all t.

Then, if $a^{-1}(\dfrac{KR}{\alpha})+c \leq B < B*$, where $a^{-1}(r)$ is the inverse function

of $a(r)$, in the region $R \times S_{B*}$ system (19.8) has a unique uniformly

asymptotically stable almost periodic solution which is bounded by

B. In particular, if $f(t,x)$ and $g(t)$ are periodic in t of

period ω, then so is the above solution.

   To prove this theorem, consider a system

$$\begin{cases} x' = f(t,x) + g(t) \\ y' = f(t,y) + g(t). \end{cases} \tag{19.9}$$

Then we have, by (ii)' and (iii)

$$\dot{V}_{(19.9)}(t,x,y) \leq \overline{\lim_{h \to 0^+}} \frac{1}{h}\{V(t+h,x+hf(t,x),y+hf(t,y))-V(t,x,y)\}$$

$$\leq -\alpha V(t,x,y),$$

and hence, Theorem 19.3 can be proved by Theorem 19.1 and the following lemma which shows the existence of a solution $\psi(t)$ of (19.8) such that $|\psi(t)| \leq B$ for $t \geq 0$.

Lemma 19.1. Under the assumptions in Theorem 19.3, if $|x_0-\phi(t_0)| \leq b^{-1}(\frac{KR}{\alpha})$, $t_0 \geq 0$, and $|x_0| \leq B$, we have

$$|x(t,t_0,x_0)-\phi(t)| \leq a^{-1}(\frac{KR}{\alpha}) \quad \text{for all} \quad t \geq t_0, \quad (19.10)$$

where $x(t,t_0,x_0)$ is a solution of (19.8) through (t_0,x_0).

Proof. Let $x(t)$ be a solution of (19.8) through (t_0,x_0), where $|x_0-\phi(t_0)| \leq b^{-1}(\frac{KR}{\alpha})$ and $|x_0| \leq B$. As long as $|x(t)| < B^*$, we have

$$V'(t,\phi(t),x(t)) \leq -\alpha V(t,\phi(t),x(t))+KR \quad (19.11)$$

by (ii)' and (iii). It follows from (19.11) that, as long as $|x(t)| < B^*$,

$$V(t,\phi(t),x(t)) \leq e^{-\alpha(t-t_0)}\{V(t_0,\phi(t_0),x_0)-\frac{KR}{\alpha}\} + \frac{KR}{\alpha}$$

$$\leq \frac{KR}{\alpha},$$

because $|x_0-\phi(t_0)| \leq b^{-1}(\frac{KR}{\alpha})$ and $V(t_0,\phi(t_0),x_0) \leq b(|\phi(t_0)-x_0|)$. Thus, by (i), as long as $|x(t)| < B^*$, $|\phi(t)-x(t)| \leq a^{-1}(\frac{KR}{\alpha})$. On the other hand, $|x(t)| \leq a^{-1}(\frac{KR}{\alpha})+c \leq B < B^*$. Therefore (19.10) is valid for all $t \geq t_0$. In particular, if $x_0 = \phi(0)$, clearly $x(t)$ is a solution of (19.8) such that $|x(t)| \leq B$.

Now we consider the case where $f(t,x)$ is linear, that is,

$f(t,x) = A(t)x$, where $A(t)$ is a continuous $n \times n$ matrix on R and is almost periodic. If the zero solution is uniformly asymptotically stable, there exists a $K \geq 1$ and an $\alpha > 0$ and a Liapunov function $V(t,x)$ defined on $I \times R^n$ such that

$$|x| \leq V(t,x) \leq K|x|,$$

$$|V(t,x_1)-V(t,x_2)| \leq K|x_1-x_2|,$$

$$\dot{V}_{(19.1)}(t,x) \leq -\alpha V(t,x).$$

If we set $W(t,x,y) = V(t,x-y)$, we have

$$|x-y| \leq W(t,x,y) \leq K|x-y|$$

and

$$|W(t,x_1,y_1)-W(t,x_2,y_2)| \leq K|(x_1-x_2)-(y_1-y_2)|.$$

Moreover, we have

$$\dot{W}_{(19.2)}(t,x,y) = \overline{\lim_{h \to 0^+}} \frac{1}{h}\{V(t+h,x+hA(t)x-y-hA(t)y)-V(t,x-y)\}$$

$$= \overline{\lim_{h \to 0^+}} \frac{1}{h}\{V(t+h,x-y+hA(t)(x-y))-V(t,x-y)\}$$

$$= \dot{V}_{(19.1)}(t,x-y) \leq -\alpha V(t,x-y) = -\alpha W(t,x,y).$$

Thus $W(t,x,y)$ satisfies the conditions in Theorem 19.3 on $I \times R^n \times R^n$. Since $x(t) \equiv 0$ is a solution of $x' = A(t)x$, c in Theorem 19.3 can be zero. By Lemma 19.1, any solution $x(t,0,x_0)$ of $x' = A(t)x+g(t)$ such that $|x_0| \leq \frac{R}{\alpha}$ satisfies $|x(t,0,x_0)| \leq \frac{KR}{\alpha}$ for all $t \geq 0$. Therefore we have the following theorem.

Theorem 19.4. Consider systems

$$x' = A(t)x \tag{19.12}$$

and

$$x' = A(t)x + g(t), \tag{19.13}$$

where $A(t)$ is a continuous $n \times n$ matrix defined on R and $g(t)$

is a continuous function on R which is bounded by M for all t.
If the zero solution of (19.12) is uniformly asymptotically stable,
then solutions of (19.13) satisfy the following properties;

(a)   There exist constants $K \geq 1$ and $\alpha > 0$ such that
$|x_0| \leq \frac{M}{\alpha}$ implies $|x(t,t_0,x_0)| \leq \frac{KM}{\alpha}$ for all $t \geq t_0$,
where $x(t,t_0,x_0)$ is a solution of (19.13).

(b)   If A(t) and g(t) are almost periodic in t, then there
is a unique almost periodic solution of (19.13) which is
bounded by $\frac{KM}{\alpha}$ and is uniformly asymptotically stable in
the large.

For functional differential equations, see [26], [79].

Concerning with the existence of a unique almost periodic solution, Nakajima has shown the following results. Now assume f(t,x)
in (19.1) is defined on R × D, where D is an open set in R^n.

Suppose that there exists a Liapunov function V(t,x,y) defined on I × D × D which satisfies the following conditions;

(i)   V(t,x,x) is bounded for $t \varepsilon I$, $x \varepsilon S$, S any compact
set in D,

(ii) $|V(t,x_1,y_1)-V(t,x_2,y_2)| \leq L\{|x_1-x_2|+|y_1-y_2|\}$ for
$t \geq 0$, $x_1,x_2,y_1,y_2 \varepsilon S$,

(iii) $\dot{V}_{(19.2)}(t,x,y) \geq a(|x-y|)$, where a(r) is continuous
and positive definite.

Moreover, assume that system (19.1) has a solution $\phi(t)$ such that
$\phi(t) \varepsilon K$ for $t \geq 0$, where K is a compact set in D. Then system
(19.1) has a unique almost periodic solution p in D and
$m(p) \subset m(f)$.

This is a special case of the following theorem which is an
improvement of a result of Fink and Seifert [20]. $V(t,x,\phi(t))$ will

be a Liapunov function in the theorem.

Theorem 19.5. Suppose that system (19.1) has a solution $\phi(t)$ such that $\phi(t) \in K$ for all $t \geq 0$, where K is a compact set in D, and assume that there exists a Liapunov function $V(t,x)$ defined on $I \times D$ which satisfies the following conditions;

(i) $V(t,\phi(t))$ is bounded on I,

(ii) $|V(t,x)-V(t,y)| \leq L|x-y|$ for $t \in I$, $x,y \in S$, where S is any compact set in D and L may depend on S,

(iii) $\dot{V}_{(19.1)}(t,x) \geq a(|x-\phi(t)|)$, where $a(r)$ is continuous and positive definite.

Then system (19.1) has a unique almost periodic solution p in D and $m(p) \subset m(f)$.

Proof. Consider a system

$$x' = g(t,x), \quad g \in H(f). \tag{19.14}$$

Then there exists a sequence $\{\tau_k\}$ such that $\tau_k \to \infty$ as $k \to \infty$ and

$f(t+\tau_k,x) \to g(t,x)$ uniformly on $R \times K$,

$\phi(t+\tau_k) \to \psi(t)$ uniformly on any compact interval in R

as $k \to \infty$. Then $\psi(t) \in K$ for all $t \in R$ and $\psi(t)$ is a solution of (19.14). If for every $g \in H(f)$, system (19.14) has only one solution which remains in K for all $t \in R$, system (19.1) has an almost periodic solution by Corollary 17.2. Thus we shall show that if system (19.14) has a solution $x(t)$ such that $x(t) \in K$ for all $t \in R$, then $x(t) \equiv \psi(t)$ for all $t \in R$.

Let $v_k(t)$ be defined by

$$v_k(t) = V(t+\tau_k, x(t)).$$

Then $v_k(t)$ is defined for $t \geq -\tau_k$ and we have

$$v_k'(t) \geq \dot{V}(t+\tau_k, x(t)) - A_k(t), \qquad (19.15)$$

where, letting $K^* \supset K$ and $L = L(K^*)$, $A_k(t) = L|f(t+\tau_k, x(t)) - g(t,x(t))|$. Clearly

$$\lim_{k\to\infty} A_k(t) = 0 \quad \text{uniformly on } R.$$

By condition (iii), we have

$$v_k'(t) \geq a(|x(t) - \phi(t+\tau_k)|) - A_k(t). \qquad (19.16)$$

For any interval $[b,c]$, if k is sufficiently large so that $b+\tau_k \geq 0$,

$$v_k(c) - v_k(b) \geq \int_b^c a(|x(s) - \phi(s+\tau_k)|)ds - \int_b^c A_k(s)ds. \qquad (19.17)$$

By conditions (i) and (ii), there exists an $M > 0$ such that

$$|v_k(c) - v_k(b)| = |V(c+\tau_k, x(c)) - V(b+\tau_k, x(b))| \leq M \quad \text{for all } k,$$

which implies that

$$\int_b^c a(|x(s) - \phi(s+\tau_k)|)ds - \int_b^c A_k(s)ds \leq M.$$

Letting $k \to \infty$, we have

$$\int_b^c a(|x(s) - \psi(s)|)ds \leq M.$$

Since b and c are arbitrary, we have

$$\int_{-\infty}^\infty a(|x(s) - \psi(s)|)ds \leq M,$$

and hence, there exist sequences $\{t_m\}$ and $\{\sigma_m\}$ such that $t_m \to -\infty$, $\sigma_m \to \infty$ as $m \to \infty$ and that $a(|x(t_m) - \psi(t_m)|) \to 0$, $a(|x(\sigma_m) - \psi(\sigma_m)|) \to 0$ as $m \to \infty$. This implies that

$$|x(t_m)-\psi(t_m)| \to 0, \quad |x(\sigma_m)-\psi(\sigma_m)| \to 0 \quad \text{as} \quad m \to \infty \qquad (19.18)$$

since $a(r)$ is continuous, positive definite and $|x(t_m)-\psi(t_m)|$, $|x(\sigma_m)-\psi(\sigma_m)|$ are bounded. In (19.17), let $b = t_m$ and $c = \sigma_m$. Then, if k is sufficiently large so that $t_m + \tau_k \geq 0$, we have

$$v_k(\sigma_m) - v_k(t_m) \geq \int_{t_m}^{\sigma_m} a(|x(s)-\phi(s+\tau_k)|)ds - \int_{t_m}^{\sigma_m} A_k(s)ds,$$

and

$$\int_{t_m}^{\sigma_m} a(|x(s)-\phi(s+\tau_k)|)ds - \int_{t_m}^{\sigma_m} A_k(s)ds - V(\sigma_m+\tau_k, \phi(\sigma_m+\tau_k))$$

$$+ V(t_m+\tau_k, \phi(t_m+\tau_k))$$

$$\leq v_k(\sigma_m) - v_k(t_m) - V(\sigma_m+\tau_k, \phi(\sigma_m+\tau_k)) + V(t_m+\tau_k, \phi(t_m+\tau_k))$$

$$\leq L\{|x(\sigma_m)-\phi(\sigma_m+\tau_k)| + |x(t_m)-\phi(t_m+\tau_k)|\}$$

$$\leq L\{|x(\sigma_m)-\psi(\sigma_m)| + |\psi(\sigma_m)-\phi(\sigma_m+\tau_k)| + |x(t_m)-\psi(t_m)|$$

$$+ |\psi(t_m)-\phi(t_m+\tau_k)|\}.$$

On the other hand, by (19.18), for each $\varepsilon > 0$ there exists an integer $N(\varepsilon) > 0$ such that if $m \geq N(\varepsilon)$,

$$|x(t_m)-\psi(t_m)| < \frac{\varepsilon}{2}, \quad |x(\sigma_m)-\psi(\sigma_m)| < \frac{\varepsilon}{2}.$$

Therefore, if $m \geq N(\varepsilon)$, letting $k \to \infty$, we have

$$\int_{t_m}^{\sigma_m} a(|x(s)-\psi(s)|)ds - \overline{\lim_{k\to\infty}}\{V(\sigma_m+\tau_k, \phi(\sigma_m+\tau_k))$$

$$- V(t_m+\tau_k, \phi(t_m+\tau_k))\} \leq \varepsilon L.$$

However, since $V(t,\phi(t))$ is bounded and $V'(t,\phi(t)) \geq 0$, $V(t,\phi(t)) \to V_0$ for some constant V_0 as $t \to \infty$, and hence, (19.19) implies that

$$\int_{t_m}^{\sigma_m} a(|x(s)-\psi(s)|) \leq \varepsilon L.$$

Letting $m \to \infty$, we have

$$\int_{-\infty}^{\infty} a(|x(s)-\psi(s)|) \leq \varepsilon L.$$

Since ε is arbitrary, we have $a(|x(t)-\psi(t)|) = 0$ or $x(t) = \psi(t)$ for all $t \in R$.

Now we shall show the uniqueness of the almost periodic solution in D. Let $\{\tau_k\}$ be a sequence such that $\tau_k \to \infty$, $f(t+\tau_k,x) \to f(t,x)$ uniformly on $R \times S$: S any compact set in D, and $\phi(t+\tau_k) \to \psi(t)$ uniformly on any compact set in R as $k \to \infty$. Then $\psi(t) \in K$ for all $t \in R$ and, as was seen above, $\psi(t)$ is the unique solution in K of system (19.1). Thus $\psi(t)$ is an almost periodic solution of system (19.1). Therefore it is sufficient to prove that $\psi(t) \equiv p(t)$ for any almost periodic solution $p(t)$ of (19.1) in D.

Suppose there exists an almost periodic solution $p(t)$ of system (19.1) such that $p(t) \in D$ for all $t \in R$ and suppose $|\psi(t_0)-p(t_0)| = \varepsilon$ at some t_0 for some $\varepsilon > 0$. Since $p(t_0) \in D$, there exists a bounded open set B such that $p(t_0) \in B \subset \bar{B} \subset D$. Since $p(t)$ is almost periodic, there exists a sequence $\{\sigma_m\}$ such that $\sigma_m \to \infty$ as $m \to \infty$ and $p(\sigma_m) \in B$ for all m. Let $v_k(t) = V(t+\tau_k,p(t))$. Then, by the same argument as used in obtaining (19.17), we have

$$v_k(\sigma_m)-v_k(t_0) \geq \int_{t_0}^{\sigma_m} a(|p(t)-\phi(t+\tau_k)|)dt - \int_{t_0}^{\sigma_m} A_k(m,t)dt, \quad (19.20)$$

where $A_k(m,t) = L_m|f(t+\tau_k,p(t))-f(t,p(t))|$ and L_m may depend on a compact set K_m of D which contains the compact set $\{p(t); t_0 \leq t \leq \sigma_m\}$. Clearly, for each fixed m

$$\lim_{k \to \infty} A_k(m,t) = 0 \quad \text{uniformly for } t \in [t_0,\sigma_m]. \quad (19.21)$$

Since $p(\sigma_m) \in B$ and we have conditions (i), (ii), there exists an

$M > 0$ such that

$$|v_k(\sigma_m) - v_k(t_0)| < M \quad \text{for all} \quad m. \qquad (19.22)$$

Letting $k \to \infty$ in (19.20), we have

$$\int_{t_0}^{\sigma_m} a(|p(t) - \psi(t)|) dt \leq M,$$

which implies

$$\int_{t_0}^{\infty} a(|p(t) - \psi(t)|) dt \leq M.$$

Since $p(t) - \psi(t)$ is almost periodic, there exists a sequence $\{t_m\}$

such that

$$|p(t_0) - \psi(t_0) - p(t_m) + \psi(t_m)| < \frac{\varepsilon}{3} \quad \text{for all} \quad m \qquad (19.24)$$

and

$$t_m \to \infty \quad \text{as} \quad m \to \infty, \; t_m + 2 < t_{m+1}. \qquad (19.25)$$

The uniform continuity of $p(t) - \psi(t)$ implies the existence of a δ,

$0 < \delta < 1$, such that

$$|p(t) - \psi(t) - p(t_m) + \psi(t_m)| < \frac{\varepsilon}{3} \quad \text{for} \quad t_m - \delta < t < t_m + \delta. \qquad (19.26)$$

From (19.24), (19.26) and $|p(t_0) - \psi(t_0)| = \varepsilon$, it follows that

$$\frac{\varepsilon}{3} < |p(t) - \psi(t)| < \frac{5\varepsilon}{3} \quad \text{for} \quad t_m - \delta < t < t_m + \delta \quad \text{and all} \quad m.$$

If we let $\min\{a(r); \frac{\varepsilon}{3} \leq r \leq \frac{5\varepsilon}{3}\} = a_0 > 0$, we have

$$\sum_{m=1}^{\infty} \int_{t_m - \delta}^{t_m + \delta} a(|p(t) - \psi(t)|) dt \geq \sum_{m=1}^{\infty} 2\delta a_0 = \infty$$

since the intervals $(t_m - \delta, t_m + \delta)$ are disjoint by (19.25). This con-

tradicts (19.23). Thus $p(t) \equiv \psi(t)$. This completes the proof.

REFERENCES

[1]   L. Amerio, Soluzioni quasi-periodiche, o limitate, di sistemi
      differenziali non lineari quasi-periodici, o limitati, Ann. Mat.
      Pura Appl., 39 (1955), 97-119.

[2]   H. A. Antosiewicz, On non-linear differential equations of the
      second order with integrable forcing term, J. London Math. Soc.,
      30 (1955), 64-67.

[3]   A. S. Besicovitch, Almost Periodic Functions, Dover, New York,
      1954.

[4]   S. Bochner, A new approach to almost periodicity, Proc. Nat. Acad.
      Sci. U. S. A., 48 (1962), 2039-2043.

[5]   L. E. J. Brouwer, Beweis des ebenen Translationssatzes, Math.
      Ann., 72 (1912), 37-54.

[6]   F. E. Browder, On a generalization of the Schauder fixed point
      theorem, Duke Math. J., 26 (1959), 291-303.

[7]   T. A. Burton, Some Liapunov theorems, SIAM J. Control, 4 (1966),
      460-465.

[8]   M. L. Cartwright, Forced oscillations in nonlinear systems, Con-
      tributions to the Theory of Nonlinear Oscillations (S. Lefschetz,
      ed.), Vol. 1, 149-241, Princeton University Press, Princeton,
      1950.

[9]   Shui-Nee Chow, Remarks on one dimensional delay-differential equa-
      tions, J. Math. Anal. Appl., 41 (1973), 426-429.

[10] Shui-Nee Chow and J. K. Hale, Strongly limit-compact maps,
      Funkcial. Ekvac., 17 (1974).

[11] Shui-Nee Chow and J. A. Yorke, Liapunov theory and perturbation
      of stable and asymptotically stable systems, (to appear).

[12] E. A. Coddington and N. Levinson, Theory of Ordinary Differential
      Equations, McGraw-Hill, New York, 1955.

[13] C. C. Conley and R. K. Miller, Asymptotic stability without uni-
      form stability: Almost periodic coefficients, J. Differential
      Eqs., $\underline{1}$ (1965), 333-336.

[14] W. A. Coppel, Almost periodic properties of ordinary differential
      equations, Ann. Mat. Pura Appl., $\underline{76}$ (1967), 27-49.

[15] C. Corduneanu, Almost Periodic Functions, Interscience Publishers,
      New York, 1968.

[16] L. G. Deysach and G. R. Sell, On the existence of almost periodic
      motions, Michigan Math. J., $\underline{12}$ (1965), 87-95.

[17] J. Favard, Lecons sur les Fonctions Presque-périodiques, Gauthier-
      Villars, Paris, 1933.

[18] A. M. Fink, Semi-separated conditions for almost periodic solu-
      tions, J. Differential Eqs., $\underline{11}$ (1972), 245-251.

[19] A. M. Fink and P. O. Frederickson, Ultimate boundedness does not
      imply almost periodicity, J. Differential Eqs., $\underline{9}$ (1971), 280-
      284.

[20] A. M. Fink and G. Seifert, Liapunov functions and almost periodic
      solutions for almost periodic systems, J. Differential Eqs.,
      $\underline{5}$ (1969), 307-313.

[21] M. Fréchet, Les fonctions asymptotiquement presque-périodiques,
      Rev. Scientifique, $\underline{79}$ (1941), 341-354.

[22] A. Halanay, Some qualitative questions in the theory of differ-
      ential equations with a delayed argument, Rev. Math. Pures
      Appl., $\underline{2}$ (1957), 127-144.

[23] _____, Asymptotic stability and small perturbations of
      periodic systems of differential equations with retarded argu-
      ments, Uspehi Mat. Nauk, $\underline{17}$ (1962), 231-233.

[24] _____, Differential Equations; Stability, Oscillation, Time
      Lags, Academic Press, New York, 1966.

[25]   A. Halanay and J. A. Yorke, Some new results and problems in the
       theory of differential-delay equations, SIAM Rev., $\underline{13}$ (1971),
       55-80.

[26]   J. K. Hale, Periodic and almost periodic solutions of functional-
       differential equations, Arch. Rational Mech. Anal., $\underline{15}$ (1964),
       289-304.

[27]   _____, Sufficient conditions for stability and instability
       of autonomous functional-differential equations, J. Differential
       Eqs., $\underline{1}$ (1965), 452-482.

[28]   _____, Ordinary Differential Equations, Wiley-Interscience,
       New York, 1969.

[29]   _____, Functional Differential Equations, Applied Math. Sci-
       ences 3, Springer-Verlag, New York, 1971.

[30]   J. K. Hale, J. P. LaSalle and M. Slemrod, Theory of a general
       class of dissipative processes, J. Math. Anal. Appl., $\underline{39}$ (1972),
       177-191.

[31]   J. K. Hale and O. Lopes, Fixed point theorems and dissipative pro-
       cesses, J. Differential Eqs., $\underline{13}$ (1973), 391-402.

[32]   J. Horvath, Sur les fonctions conjuguées a plusieurs variables,
       Nedel. Akad. Weten Proc., Ser. A., $\underline{56}$ (1953), 17-29.

[33]   G. S. Jones, Asymptotic fixed point theorem and periodic systems
       of functional-differential equations, Contrib. Differential
       Eqs., $\underline{2}$ (1963), 385-405.

[34]   J. Kato, Uniformly asymptotic stability and total stability,
       Tohoku Math. J., $\underline{22}$ (1970), 254-269.

[35]   J. Kato and A. Strauss, On the global existence of solutions and
       Liapunov functions, Ann. Mat. Pura Appl., $\underline{77}$ (1967), 303-316.

[36]   J. Kato and T. Yoshizawa, A relationship between uniformly asy-
       mptotic stability and total stability, Funkcial. Ekvac., $\underline{12}$
       (1970), 233-238.

[37] J. P. LaSalle, A study of synchronous asymptotic stability, Ann.
      Math., 65 (1957), 571-581.

[38] _____, The extend of asymptotic stability, Proc. Nat. Acad.
      Sci. U. S. A., 46 (1960), 363-365.

[39] _____, Asymptotic stability criteria, Proc. of Symposia in
      Appl. Math., 13 (1962), 299-307.

[40] _____, Stability theory of ordinary differential equations,
      J. Differential Eqs., 4 (1968), 57-65.

[41] J. P. LaSalle and S. Lefschetz, Stability by Liapunov's Direct
      Method with Applications, Academic Press, New York, 1961.

[42] S. Lefschetz, Differential Equations; Geometric Theory, 2nd Ed.,
      Interscience, New York, 1963.

[43] J. J. Levin, On the global asymptotic behavior of nonlinear sys-
      tems of differential equations, Arch. Rational Mech. Anal.,
      5 (1960), 194-211.

[44] J. J. Levin and J. A. Nohel, Global asymptotic stability for non-
      linear systems of differential equations and applications to
      reactor dynamics, Arch. Rational Mech. Anal., 5 (1960), 194-211.

[45] J. L. Massera, On Liapunoff's conditions of stability, Ann. Math.,
      50 (1949), 705-721.

[46] _____, The existence of periodic solutions of systems of
      differential equations, Duke Math. J., 17 (1950), 457-475.

[47] _____, Contributions to stability theory, Ann. Math., 64
      (1956), 182-206.

[48] _____, Erratum: Contributions to stability theory, Ann.
      Math., 68 (1958), 202.

[49] R. K. Miller, Asymptotic behavior of solutions of nonlinear dif-
      ferential equations, Trans. Amer. Math. Soc., 115 (1965),
      400-416.

[50]   _____, Asymptotic behavior of nonlinear delay-differential
        equations, J. Differential Eqs., $\underline{1}$ (1965), 293-305.

[51]   _____, Almost periodic differential equations as dynamical
        systems with applications to the existence of a. p. solutions,
        J. Differential Eqs., $\underline{1}$ (1965), 337-345.

[52]   T. Naito, On the uniqueness in the hull, Tohoku Math. J., $\underline{25}$
        (1973), 383-389.

[53]   F. Nakajima, Existence of quasi-periodic solutions of quasi-
        periodic systems, Funkcial. Ekvac., $\underline{15}$ (1972), 61-73.

[54]   _____, Separation condition and stability properties in al-
        most periodic systems, Tohoku Math. J., $\underline{26}$ (1974).

[55]   C. Olech and Z. Opial, Sur une inégalité différentielle, Ann.
        Polon. Math., $\underline{7}$ (1960), 247-254.

[56]   N. Onuchic, Invariance properties in the theory of ordinary dif-
        ferential equations with applications to stability problems,
        SIAM J. Control, $\underline{9}$ (1971), 97-104.

[57]   Z. Opial, Sur une équation différentielle presque-périodique sans
        solution presque-périodique, Bull. Acad. Polon. Sci. Ser. Sci.
        Math. Astron. Phys., $\underline{9}$ (1961), 673-676.

[58]   N. Pavel, On dissipative system, Boll. Unione Mat. Italiana,
        $\underline{4}$ (1971), 701-707.

[59]   R. Reissig, Zur Theorie der erzwungenen Schwingungen, Math.
        Nachr., $\underline{13}$ (1955), 309-312.

[60]   _____, Ein Beschränktheitssatz für gewisse nichtlineare
        Differentialgleichungen dritter Ordnung, Mber. Dt. Akad. Wiss.,
        $\underline{6}$ (1964), 481-484.

[61]   G. Seifert, Almost periodic solutions for almost periodic systems
        of ordinary differential equations, J. Differential Eqs., $\underline{2}$
        (1966), 305-319.

[62] _____, Almost periodic solutions and asymptotic stability,
J. Math. Anal. Appl., 21 (1968), 136-149.

[63] G. R. Sell, Periodic solutions and asymptotic stability, J.
Differential Eqs., 2 (1966), 143-157.

[64] _____, Nonautonomous differential equations and topological
dynamics, I, II, Trans. Math. Soc., 127 (1967), 241-262, 263-
283.

[65] A. Strauss and J. A. Yorke, On asymptotically autonomous differential equations, Math. Systems Theory, 1 (1967), 175-182.

[66] _____, Perturbing uniform asymptotically stable nonlinear
systems, J. Differential Eqs., 6 (1969), 452-483.

[67] K. E. Swick, On the boundedness and the stability of solutions
of some non-autonomous differential equations of the third
order, J. London Math. Soc., 44 (1969), 347-359.

[68] _____, Invariant sets and convergence of solutions of nonlinear differential equations, J. Differential Eqs., 10 (1971),
204-218.

[69] L. H. Thurston and J. S. W. Wong, On global asymptotic stability
of certain second order differential equations with integrable
forcing terms, SIAM J. Appl. Math., 24 (1973), 50-61.

[70] I. Vrkov, Integral stability, Czech. Math. J., 9 (1956), 71-128.

[71] D. R. Wakeman, An application of topological dynamics to obtain
a new invariance property for nonautonomous ordinary differential equations, J. Differential Eqs., (to appear).

[72] J. A. Yorke, Invariance for ordinary differential equations,
Math. Systems Theory, 1 (1967), 353-372.

[73] T. Yoshizawa, Liapunov's function and boundedness of solutions,
Funkcial. Ekvac., 2 (1959), 95-142.

[74] _____, Existence of a bounded solution and existence of a
periodic solution of the differential equation of the second

order, Mem. Coll. Sci. Univ. Kyoto, Ser. A, <u>33</u> (1960), 301-308.

[75]   _____, Asymptotic system of a perturbed system, Proc.
International Symp. on Nonlinear Differential Eqs. and Mechanics,
Colorado Springs, 1961, 80-85.

[76]   _____, Asymptotic behavior of solutions of a system of differential equations, Contrib. Differential Eqs., <u>1</u> (1963),
371-387.

[77]   _____, Stable sets and periodic solutions in a perturbed
system, Contrib. Differential Eqs., <u>2</u> (1963), 407-420.

[78]   _____, Ultimate boundedness of solutions and periodic solution of functional-differential equations, Colloques Inter-
nationaux sur les Vibrations Forcées dans les Systèmes Non-
linéaires, Marseille, Sept., 1964, 167-179.

[79]   _____, Extreme stability and almost periodic solutions of
functional-differential equations, Arch. Rational Mech. Anal.,
<u>17</u> (1964), 148-170.

[80]   _____, <u>Stability Theory by Liapunov's Second Method</u>, The
Mathematical Society of Japan, Tokyo, 1966.

[81]   _____, Existence of a globally uniform-asymptotically stable
periodic and almost periodic solution, Tohoku Math. J., <u>19</u>
(1967), 423-428.

[82]   _____, Stability and existence of a periodic solution, J.
Differential Eqs., <u>4</u> (1968), 121-129.

[83]   _____, Some remarks on the existence and the stability of
almost periodic solutions, Studies in Applied Mathematics,
<u>5</u> (1969), 166-172.

[84]   _____, The existence of almost periodic solutions of functional-differential equations, Proc. International Conference
on Nonlinear Oscillations, Kiev, 1969.

[85]   _____, Asymptotically almost periodic solutions of an al-

    most periodic system, Funkcial, Ekvac., 12 (1969), 23-40.

INDEX